A Flower A Day

A Flower A Day

Miranda Janatka

BATSFORD

First published in the United Kingdom in
2022 by
Batsford
43 Great Ormond Street
London WC1N 3HZ
An imprint of B.T. Batsford Limited

ISBN: 9781849947176

A CIP catalogue record for this book is
available from the British Library.

30 29 28 27 26 25 24
10 9 8 7 6 5 4 3

Reproduction by Rival Colour Ltd, UK
Printed by Leo Paper Products, China.

This book can be ordered direct
from the publisher at the website:
www.batsfordbooks.com

Disclaimer: All plants are potentially toxic.
Seek the identification and advice of a
professional before using any as a food
or herbal remedy. This book documents
usage and is not intended to advise on any
consumption or application.

MIX
Paper | Supporting
responsible forestry
FSC
www.fsc.org FSC® C020056

Previous page:
The unfurling
flowers of Lacy
phacelia (*Phacelia
tanacetifolia*) are
attractive to bees
and other insects.

Below: Common
red poppies
(*Papaver rhoeas*)
growing through
a field of wheat
in Friuli Venezia
Giulia, Italy.

CONTENTS

Introduction 6

JANUARY 12

FEBRUARY 42

MARCH 70

APRIL 100

MAY 132

JUNE 164

JULY 194

AUGUST 224

SEPTEMBER 254

OCTOBER 285

NOVEMBER 316

DECEMBER 346

Index 378

Picture Credits 383

Acknowlegements 384

INTRODUCTION

Right: Flowers have long been used by artists to capture and represent the beauty of nature. *Bloemen in Blauwe Vaas* by Vincent van Gogh, oil on canvas (1887).

Whether you're curious about plants and science, or someone with an interest in art, humankind has long been deeply connected with, and drawn to, flowers. Wrapped up with our cultural events from birth to death, flowers have not only represented worship, love and remembrance – as well as the brevity of life – but have also had a very real connection to our survival. The occurrence of blooms is suggestive of fertile earth and the possibility of food to come. The delight people find today in flowers may well stem from the relief of a return of spring, a period of rain or quite literally a forthcoming fruitful season. Also, that flowers are often used as national emblems or symbols, signifies their importance to the cultural and historical identity of nations. Pride in a particular plant that has led to an industrial production profiting the land or one that is native and perhaps found only in that location, becomes of significant value to those who live there.

We experience flowers either as a commercial product (such as bouquets of flowers), out in the wild or as the result of our own intimate connection with nature, nurturing and growing them ourselves. Research has shown that the gift of flowers makes people more likely to smile and create increased social contact, but we also know that being out in nature and even just observing plants increases our well-being. One reason for this is that the sight of fractals found in plants – infinitely complex patterns that repeat simple processes – reduces stress levels (an example of a fractal would be the arrangement of many small flowers within an inflorescence forming a Fibonacci spiral). However, flowers in particular offer more than just a visual experience; they are multi-sensory with their movement, texture, scent and occasionally taste, which delight and can tap into our long-term memories, connecting us emotionally to our autobiographical past, as well as to each other. In addition, for gardeners, a flowering plant displaying blooms suggests that it is at the climax of its growth; the flowers may represent success and maturity, a sign of the plant completing its first – and possibly only – life cycle.

Lastly, we cannot ignore the importance of flowers to insects and animals beyond ourselves, as well as the greater ecosystem. Operating as a map of evolution, one can trace through the transformation

Above: *Dahlia* 'Blyton Softer Gleam' is a small ball-type dahlia that lasts well as a cut flower in a vase.

Opposite: Helenium is also known as sneezeweed as it was used for making snuff. *Helenium* 'Rubinzwerg' is shown here.

of flowers, their relationship with the pollinators they feed (or pretend to). From the tailored attraction of scent and colour aimed at particular insects and birds, to sophisticated plant mimicry (such as that of a bee orchid, which literally resembles the female bee to attract the male). While all parts of plants have evolved back and forth to adapt to changing conditions in order to survive, flowers are the point of most visual difference, which is why they were the basis for the first classification used to key out and name plants. While plants today are increasingly classified using their DNA, plant identity and classification was first worked out using floral morphology. The modern use of this started with Carl Linnaeus's adoption of binomial names for plants in his *Species Plantarum* of 1753, using botanical nomenclature (the two-part names of plants that we see in Latin).

The flowers in this book have been selected to represent some of the most incredible plants from around the world, most of which are easily happened upon on nature walks or in the gardens of temperate regions. The plants chosen include the largest to the smallest blooms in the world, the most commercially valuable and those that are used again and again in literature and art to convey either secret or overt meanings. The intention of this book is to enhance an enjoyment of discovering flowers as they unfold throughout the year, as well as transporting the mind to faraway places from the comfort of one's home. Hearing the stories of flowers increases their significance to us: they are nature's own art, helping craft our own narratives, personal and as a collective. This book may help influence your choice of plants with which you choose to grow and decorate your home and garden, or to gift a loved one with the knowledge of their secret language. For myself, appreciation of any art form cannot help but hark back to the natural world, and flowers provide much of the origin of beauty found around us today.

Each flower in this book has been chosen to illustrate a particular day, and while flowering times will vary somewhat from year to year, with some flowers blooming over a longer period of time than others, they have each been selected to indicate a time in bloom of the region most heavily referenced to in the entry. This is often the area of origin, but where no specific region is mentioned, this will generally correlate to the time the plant flowers in the UK.

By examining some of the uses, benefits and histories of each plant, sharing their stories and revealing their relationship to our human world, a floral realm is opened up for discovery.

Under each date the common name for each flower is given, as well as the most recent botanical name, in italics as is the convention. As set out by Carl Linnaeus, each botanical name is made up of two parts: first the genus and second the species. To describe a species in the plural, the abbreviation 'spp.' is used, and should a cultivar be named, this is given at the end of the botanical name designated by single quotation marks (for example, *Iris* 'Katharine Hodgson'). Each plant also has a family name, which, to keep things simple, has been

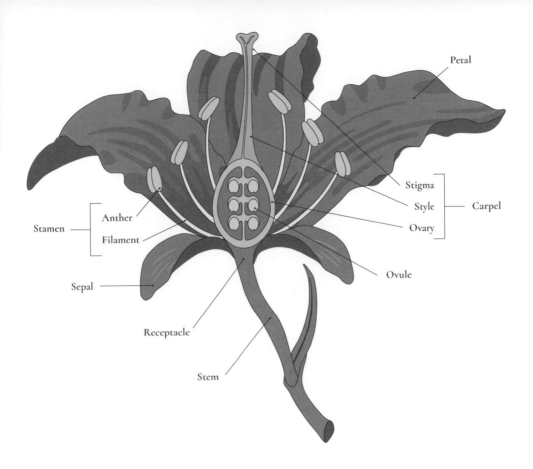

Petal

Stigma

Style

Carpel

Ovary

Ovule

Anther

Stamen

Filament

Sepal

Receptacle

Stem

left out of the texts, but these do also further connect the flowers to each other and may be referred to at certain points in the book. For example, plants within the mint family (*Lamiaceae*) will share certain characteristics, although often much more broadly than those within the same genus.

Only a very simple understanding of flowering parts is required while reading this book. The illustration above provides a quick reminder of the names of some of the basic parts of a flower.

The beauty and sophistication of flowers may offer a gateway to a lifelong love of plants. Species new to modern science are being discovered all the time, and with them the potential benefits they offer us, as much-needed new medicines or sources of food. Certain flowers will also hold a very special place in our hearts for many reasons, perhaps deeply personal. I grow gladioli in remembrance and celebration of my father who admired them, and this brings me a great sense of joy and connection to him each year on the days they bloom.

I hope that this book offers you both insight and an enjoyable seasonal tour through some of the most incredible and evocative details of our natural and very human world.

Above: Anatomy of a flower illustrating and identifying the basic parts.

Opposite: Gladioli are classic garden plants, flowering typically from June until October each year.

COMMON SNOWDROP
Galanthus nivalis

Common snowdrop flowers can be spotted as early as the very first day of the year.

These dainty flowers are seen as a symbol of hope, being one of the very first blooms of the year. They are often seen poking up through the snow, suggesting that winter is slowly drawing to a close. Many cultivars are bred by keen collectors, who may spend significant amounts of money purchasing unusual varieties, before themselves breeding flowers with different but subtle markings in green, yellow and peach. These colours are mostly found on the underside of the flower, and it's for this reason that it is said the best way to view snowdrops is from the ground, looking up.

CLEMATIS 'WINTER BEAUTY'
Clematis urophylla 'Winter Beauty'

The many blooms of clematis 'Winter Beauty' are a delight to see as early as January.

This valuable garden plant provides green foliage all year round as well as producing many attractive, bell-shaped flowers at a time of year when not many other blooms are to be found. The broad leaves and abundance of large flowers could fool you into believing it is midsummer rather than winter. As a vigorous climber, 'Winter Beauty' is used in gardens to provide both ornamental value and a source of food for early emerging honey bees, and it will scale fences, shrubs and even trees.

NIGHT-SCENTED PELARGONIUM
Pelargonium triste

Night-scented
pelargonium
flowers during
the spring and
early summer
months in
South Africa.

This, one of the first pelargoniums to be collected from the wild and then cultivated around the world, was brought to England in 1632 from South Africa. In the plant trade, pelargoniums are often mistakenly called geraniums; however, they are quite different plants. Unlike geraniums, pelargoniums are not hardy to frost, so need to be kept indoors during the cooler months in colder countries. You can tell pelargoniums and geraniums apart by studying the flowers: the top two petals of a pelargonium will be shaped slightly differently to the bottom three, unlike a geranium which has a strictly symmetrical appearance.

TEA PLANT
Camellia sinensis

Legend has it that Chinese farmers would train and use monkeys to gather tea leaves as shown in this copperplate engraving, 1821.

This evergreen shrub is used to produce black, white, yellow and green tea. It is believed that the mythical Chinese ruler Shennong created the drink around 5,000 years ago. He was rumoured to have been drinking a cup of hot water under a camellia plant when a leaf fell into his cup. He left it to brew and reportedly enjoyed the taste. Originally a drink for the wealthy due to its high cost, the East India Company played a major role in importing tea into the UK in large quantities in the 17th century, helping to make it more affordable to the wider public.

WINTER-FLOWERING HONEYSUCKLE
Lonicera fragrantissima

A delight to see in January, winter-flowering honeysuckle provides a much-needed food source for insects.

This flower was introduced to the UK from China in 1845 and to the USA a few years later. It was grown in gardens and around the doors of homes in Victorian times to ward off evil spirits and witches. The plant itself benefits many creatures, attracting the elephant hawk-moth, which is preyed upon by bats. The climbing stems provide nest sites for birds and the new shoots attract blackfly, which feed ladybirds and lacewings.

NERINE/BOWDEN LILY
Nerine bowdenii

Growing wild in South Africa, the Bowden lily flowers from the summer into the autumn months.

Named after the sea nymphs of Greek legend, the Nereids, these flowers are found in the mountain scree of South Africa and were first introduced to Britain at the beginning of the 20th century. The flowers are produced after the leaves die down, creating a colourful display, and they are a provider of nectar late in the year for insects. Nerines are believed to represent freedom and good fortune.

YELLOW SAGE/COMMON LANTANA
Lantana camara

In Central and South America where common lantana is native, it blooms frequently and can be seen in flower this month.

Native to South America, this plant is now popular in gardens in temperate countries and cultivated indoors where temperatures are not naturally warm enough. It was first brought to Europe by Dutch explorers, and later to Asia where unfortunately it grows too happily, having since established itself as a weed. Studies have shown the leaves can contain antimicrobial, fungicidal and insecticidal properties, and they have certainly been used in traditional medicine for the treatment of illnesses including measles and chickenpox.

CHILEAN LANTERN TREE
Crinodendron hookerianum

Growing in its home of native Chile, the buds of the Chilean lantern tree form in autumn and then flower in the summer months.

From Chile, as the name suggests, this evergreen shrub produces many lantern-shaped, crimson flowers suspended on long stalks. Popular as an exotic-looking plant that can tolerate cold weather in sheltered spots, it thrives in the UK and grows best in slightly acidic soils. The botanical name combines the Greek words *krinon* meaning 'lily' and *dendron* meaning 'tree'. In Chile, it is used in traditional medicine to induce vomiting to expel toxins from the body.

NIGHT-SCENTED PHLOX
Zaluzianskya ovata

Night-scented phlox fills the night air of southern Africa with perfume, where it grows wild in the summer months.

Grown for its intense fragrance, which is particularly noticeable at night, this plant is native to southern Africa. During the day, the red buds are visible, and these unfurl slowly to reveal the white flowers in the evening, releasing a strong and spicy fragrance. It is popular with gardeners around the world, especially grown in pots during the warmer months in cooler climates.

HOLY BASIL
Ocimum tenuiflorum

Holy basil is widespread in the Indian subcontinent and used alongside turmeric as part of Indian Ayurvedic medicine.

Closely related to common basil, this plant plays an important role in Hinduism, where it is known as *tulsi* and is said to be a manifestation of the goddess of the same name. Lord Krishna is said to wear a garland of the leaves and flowers around his neck, with the plant believed to protect and purify. The leaves are edible and have a pungent flavour, which is reminiscent of clove, mint and basil, and is intensified when cooked.

FALLING STARS
Crocosmia aurea

This illustration by Walter Fitch, from *Curtis's Botanical Magazine* (1847), shows the flowers of falling stars, which bloom in South Africa's summer months.

Found in South Africa around stream banks and forest margins, the name 'crocosmia' comes from the Greek words *krokos* meaning 'saffron' and *osme* meaning 'smell'. It is said that the dried flowers placed in warm water create a saffron-like scent. A member of the iris family, crocosmia are also known as montbretia. In the wild, the plant provides food for birds, which eat the seeds after the flowers have gone over, while bush pigs feed on the corms. The tall stems make the plant attractive in a vase as a cut flower.

BANANA
Musa acuminata

Musa acuminata, known much more commonly to us as banana, can produce large flowers all year round in tropical countries or inside large, heated greenhouses.

Parent to many of the modern edible dessert bananas now eaten around the world (including the most widely consumed banana cultivar, 'Dwarf Cavendish'), *Musa acuminata* is native to southern Asia and was first cultivated by humans around 8000 BCE. Each flower forms part of a group (known as an inflorescence) and grows horizontally from the trunk. The female flowers are found near the base and develop into fruit, while the male flowers are located higher up and do not. The plant is also grown for ornamental purposes, including as a houseplant in cooler parts of the world.

BACOPA
Chaenostoma cordatum

Originating from South Africa, along both coastal stretches as well as forested kloofs (valleys), the name *Chaenostoma* is derived from the Greek meaning 'gaping mouth' referring to the wide, open centre of the small, star-shaped flowers. The word *cordatum* is from Latin, referring to the heart-shaped leaves. It is a tender plant, used in hanging baskets for summer growing in temperate countries, producing many cascading flowers. After flowering, the fruit is produced in the form of a capsule containing amber-coloured seeds.

14TH JANUARY

WINTER JASMINE
Jasminum nudiflorum

Native to China, this flower is now widely cultivated and popular due to its abundance of flowers produced on bare stems before leaves form each year. Known in China as the flower that welcomes spring, it provides an abundance of foliage, which is often used to cover walls in gardens. The scentless flowers have been admired in the West since the Scottish botanist Robert Fortune first brought the plant over to Britain from China in 1844.

Right: The African lily which flowers in South Africa's late summer, is depicted here in a hand-coloured copperplate engraving by George Cooke (1817).

Opposite top: While the flowers of *Chaenostoma cordatum* are white, bred cultivars produce flowers in pink and purple such as bacopa 'Gulliver Violet' which are popular for ornamental displays and bloom all year round in South Africa.

Opposite bottom: Popular and reliable for cheery colour in the darker months, winter jasmine starts flowering from January through to March in British gardens.

Agapanthus minor.

AFRICAN LILY / LILY OF THE NILE
Agapanthus africanus

Seen as a symbol of love, the Latin name of this flower comes from the Greek word *agape* meaning 'love' and *anthos* meaning 'flower'. It is native to South Africa, where it is regarded as an aphrodisiac, and has been worn by women for strength and to boost fertility. It is also believed to protect against unwanted thunderstorms. African lilies are grown in gardens around the world as their large blue-headed flowers provide impressive displays throughout warmer months.

ORANGE WITCH HAZEL
Hamamelis × intermedia 'Jelena'

A blue tit lands on the winter blooming branches of witch hazel 'Jelena', which produces orange blooms from January and throughout February.

This is one of the most popular witch-hazel cultivars, producing an abundance of orange flowers, rather than the more commonly seen yellow blooms. The flowers curl like narrow strips of fruit peel, releasing a strong citrussy scent. It was bred in the 1950s by Belgian plant breeder Robert de Belder, and named for his wife, who was a highly regarded botanist and horticulturist in her own right. It was the recognition of this plant that led them to design what would later become a world-famous botanical garden, Arboretum Kalmthout.

CORNELIAN CHERRY
Cornus mas

This engraving by Jean Matheus in Renouard's translation of Ovid's *Metamorphoses*, shows Circe with a dish of Cornelian cherries, having turned Odysseus's followers into pigs (c.1610).

Native to southern Europe and southwestern Asia, this deciduous tree produces small yellow flowers in late winter. The flowers are produced before the leaves emerge, which make them particularly attractive in gardens. It was first introduced to the West during the Middle Ages where it was grown in monastic gardens. Following the flowers are red fruit, which are very bitter until they are completely ripe, by which time they taste like plums. The plant is mentioned in various works of classical literature, including Homer's *Odyssey*, where the followers of Odysseus are fed with the fruits of the tree after they are transformed into pigs.

Iris unguicularis, also known as the winter iris, defies the cold and can be found in bloom this month, making it a valuable plant for gardeners.

ALGERIAN IRIS
Iris unguicularis

This winter-flowering iris is faintly and sweetly scented with delicate markings on its lavender petals. Native to Greece, Turkey and some surrounding areas, the flowers appear for a few months from midwinter, making them desirable with gardeners as early-flowering plants. The cultivar *Iris unguicularis* 'Mary Barnard' is particularly popular, winning awards for being especially reliable and having a darker purple colour to the blooms.

Wintersweet
produces a heady
fragrance in
winter gardens
to attract
pollinators.

WINTERSWEET
Chimonanthus praecox

Native to China, this is another popular garden plant as it provides scented flowers in midwinter, before leaves have started to emerge. Brought over to Britain from China in 1766, Lord Coventry planted it in his conservatory at Croome Court, as it was only 100 years later that it was understood to be a winter-hardy plant. Evidently a hit, the so-called 'father of English gardens' John Loudon proclaimed that 'No garden should be without it.' The scent has been described as a mix of jonquil-type daffodils and violets. As with other heavily scented sweet flowers, the smell is almost unpleasant in excess, so should really be used sparingly when brought inside the home.

OREGON GRAPE 'WINTER SUN'
Mahonia × media 'Winter Sun'

The evergreen shrub Oregon grape 'Winter Sun' produces many bold and bright flowers in mid-winter.

A plant bred by crossing *Mahonia oiwakensis* subsp. *lomariifolia* and *Mahonia japonica*, this cultivar is favoured for producing flowers earlier in the year. The fragrant flowers that bloom on arching stems are followed by purple berries into the summer months. As an evergreen plant, it is grown to add architectural structure to a garden, and with spiny leaves reminiscent of a holly, it is also used to discourage trespassers.

EVERGREEN CLEMATIS
Clematis cirrhosa

The evergreen clematis produces many large blooms in the winter months. 'Freckles', shown, is speckled with maroon freckles.

Popular with gardeners as it flowers through the winter months, this plant is native to the Mediterranean. It is a climber with bell-shaped flowers, followed by silky seed heads once the blooms have gone over. This plant has popular cultivars, such as 'Wisley Cream' named after the famous garden in which it was raised: the Royal Horticultural Society Garden Wisley in Surrey, England. Another common name is Early Virgin's Bower, which refers to the habit of the abundantly produced flowers to cascade downwards, the colour white being a reference to virginity.

MĀNUKA/MANUKA

Leptospermum scoparium

Manuka flowers for only a few weeks at the beginning of summer in New Zealand, with each individual bloom opening for just a few days.

Honey has been used as a traditional medicine for over 2,000 years, with the creamier and slightly nutty-tasting Manuka honey regarded as the most effective. The Manuka bush is commonly found across the coasts of the North and South Islands of New Zealand and parts of Australia, where it is believed to have originated. The Maori people of New Zealand have long recognized the healing properties of this plant. The nectar of the flowers contains high levels of methylglyoxal and phenols, known for their particularly effective antibacterial and antiseptic qualities.

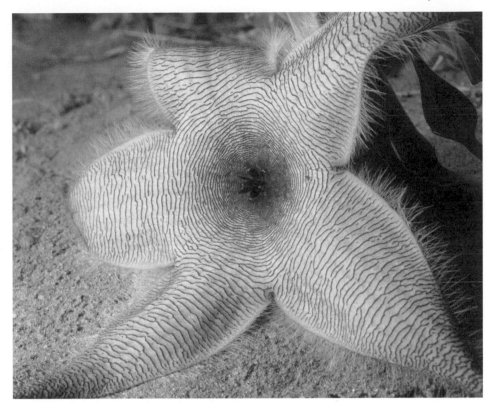

CARRION FLOWER
Stapelia gigantea

The starfish-shaped blooms of the carrion flower can be spotted in sandy and rocky areas during South Africa's spring and summer months.

This flowering plant is native to desert regions of South Africa and Tanzania, and popular with many collectors of succulent plants. The flowers are large and star-shaped with a silky texture; they exude a smell of rotting flesh in order to attract the flies that pollinate it and can grow up to 40cm (16in) in diameter. It is believed the plant was traditionally used to treat what was historically termed 'hysteria'.

The winter-flowering plum blossom is shown here in this colour woodblock mokuhanga, from the Japanese edition of Wang Gai's *Jieziyuan Huazhuan* or *Mustard Seed Garden Painting Manual* (1812).

PLUM
Prunus mume

Also known as Japanese apricot, which is a more accurate description of the fruit it produces than 'plum', this deciduous tree is grown more for ornamental purposes than for the fruit itself. In mid- to late winter it puts on a display of pink flowers that have a slightly spicy fragrance. The flowers are followed by apricot-like fruit, but unlike the species grown for eating, these are bitter and the stone inside clings to the flesh. The flower has long been depicted in Chinese art and poetry, symbolizing both winter and the coming of spring.

PURPUS HONEYSUCKLE
Lonicera × purpusii

One of the most popular winter-flowering plants for gardeners, fragrant blooms of purpus honeysuckle will flower until March.

This hybrid plant was created from two Chinese species, *Lonicera fragrantissima* and *Lonicera standishii* in Germany in the 1920s. The flowers have a strong, characteristic honeysuckle scent. Popular with gardeners as it can tolerate cold weather and flowers in the middle of winter, the cultivar 'Winter Beauty' is particularly popular. The tubular flowers are produced before the leaves emerge, showing on the bare branches. Unlike many honeysuckles, this plant grows as a shrub with an arching form, rather than as a climber.

STINKING HELLEBORE
Helleborus foetidus

This elegant native flower, the stinking hellebore, will start to flower from this month into early spring.

Native to the UK, as well as parts of central and southern Europe, this plant is poisonous so care must be taken handling it. While the flowers are not particularly foul smelling, crushing the leaves produces an odour sometimes compared to beef. The plant is evergreen and the flowers a yellowish green. The many stamens make it a valuable source of nectar for bees and other pollinators at this time of year. It has been found that yeast inside the flowers helps the plant attract pollinators by raising the temperature and therefore increasing the evaporation of scented compounds into the air.

Right: Narcissus shown here admiring his own reflection in a hand-coloured halftone reproduction of a 19th-century illustration by an unknown artist.

Opposite top: The bright red buds of the Persian ironwood grace its bare branches from January and into February.

Opposite bottom: *Daphne bholua* 'Jacqueline Postill' produces clusters of fragrant flowers for a few weeks in the middle of winter.

DAFFODIL 'RIJNVELD'S EARLY SENSATION'
Narcissus 'Rijnveld's Early Sensation'

If you spot a compact, bold yellow daffodil out in bloom this early in the year, there's a good chance it is this cultivar. A pure, golden colour, 'Rijnveld's Early Sensation' is popular with gardeners for bringing colour to the winter garden. With a slight fragrance, this flower is also able to tolerate the snow, making it a reliable favourite. In Greek mythology, Narcissus, whom some believe daffodils are named after as a reference to their nodding flowers, was the son of the river god Cephissus and the nymph Liriope. He was known for his beauty and fell in love with his own reflection in a pool of water; as this was unobtainable, Narcissus pined away.

PERSIAN IRONWOOD
Parrotia persica

A relative to witch hazel, this plant is native to northern Iran. During this month and the next, the flowers bloom in crimson clusters at the tips of bare branches. The name 'ironwood' refers to the tree's extremely dense wood. As well as the flowers, the flaking bark makes an attractive feature in winter. The flowers themselves do not have petals; it is the flower stamens emerging from the buds that produce the vibrant and attractive colour.

NEPALESE PAPER PLANT
Daphne bholua

A popular winter-flowering plant, with the cultivar 'Jacqueline Postill' a particular favourite in gardens as it always flowers in January. The flowers are typical of other daphnes and produce a sweet, citrus-like scent. Native to China and Nepal, as the common name suggests, the flowers are rich in nectar, popular with bees and followed by very dark purple berries.

GLAUCOUS SCORPION-VETCH
Coronilla valentina subsp. *glauca*

Common brimstone butterflies hibernate over winter but can still be spotted feeding on the nectar of flowers such as the glaucous scorpion-vetch on warm, sunny days.

Heavily fragrant and with a scent resembling peaches – yellow, pea-like blooms adorn this plant from late January into spring. Native to the Mediterranean, the cultivar 'Citrina' with lemon-yellow flowers is more commonly grown in gardens. The genus name *Coronilla* refers to the way in which the flowers are produced in a rounded shape, like a crown. The subspecies *glauca*, as the name suggests, is a plant with grey-green foliage as opposed the greener foliage of *Coronilla valentina*.

COMMON CHICKWEED
Stellaria media

Chickweed is a low-growing plant that is rich in minerals and makes a nutritious snack for chickens. Humans can eat it too, in salads.

Native to Eurasia, this plant is now found throughout much of the world and regarded as a weed. It is edible and nutritious, eaten by both birds and humans (although it is not recommended for those pregnant or breastfeeding). The white flowers comprise five sets of petals that resemble rabbit ears. The genus name *Stellaria* refers to the flower's star-shape form. In traditional medicine it has been used to treat skin conditions and arthritis alongside many other ailments.

NARRA
Pterocarpus indicus

Blooms of the narra tree will start to open from the end of this month in tropical and temperate parts of Asia, with the yellow flowers lighting up the branches.

This national tree of the Philippines is believed to symbolize strength of character and strong spirit. It has yellow flowers arranged on long stalks, and its attractive form is why the tree is used for ornamental purposes, to line avenues. The wood of the tree is hard and rose-scented, used for high-grade furniture and decorative veneer. Infusions of the leaves are used in shampoos, as well as in traditional medicine to treat a variety of health problems, including throat ailments.

BELL HEATHER
Erica cinerea

The flowers of bell heather appear from late autumn through to late spring and can be found growing in UK heathlands.

Native to western Europe, bell heather is found most abundantly in places such as Britain and Ireland, and particularly in harsh habitats, such as the moorlands and heathlands of Scotland. It produces pollen prolifically and is important for bees like the buff-tailed and red-tailed bumblebee and honey bees. It is a source of heather honey, a smoky and mildly sweet honey with a long-lasting taste. It is dark amber in colour and prized for being rich in antioxidants and for its anti-bacterial properties.

HOLLYHOCK
Alcea rosea

Native to Turkey, this species is the parent of the many popular and tough-growing garden hollyhock plants. Traditionally, children made dolls out of the flowers by using the fully-opened flowers as skirts, the half-opened flowers as torsos and the buds for heads, held together with a stick. Cultivated all over the world, hollyhocks are attractive to butterflies and hummingbirds, and are also known as 'outhouse flowers' or 'privy plants' as their tall stems and large blooms were traditionally used to hide unsightly buildings such as toilets.

Hollyhocks are also native to China, where, starting this month, they will flower through until August.

EARLY-FLOWERING BORAGE
Trachystemon orientalis

Also known as Abraham-Isaac-Jacob, buds of the early-flowering borage will just be starting to make an appearance before opening up into star-shaped blooms.

S tarting this month, this relative of common borage starts to display blooms of bluish-purple flowers with white throats. A tough plant, which is quick to spread, it is native to southern Europe and southwestern Asia. The pointed flowers bloom as the foliage is just beginning to develop. The genus name *Trachystemon* comes from the Greek *trachys* meaning 'rough', and *stemon* in reference to the filaments found on the flowers.

NASTURTIUM
Tropaeolum majus

Stylized illustration of nasturtium from the *Systematischer Bilder-Atlas zum Conversations-Lexikon Ikonographische Encyklopädie* (1875).

Native to South and Central America, nasturtiums are popular in gardens as they grow quickly, and are showy with bright flowers. The genus name *Tropaeolum* comes from the Greek *tropaion* (and the Latin *tropaeum*), meaning 'trophy'. The Romans would hang the armour and weapons of their vanquished enemies on a trophy pole, and it was believed that the rounded leaves of nasturtium were thought to look like shields, and the orange flowers like blood-stained helmets, thus resembling these 'trophies'. All parts of the plant are edible and have a pleasant, peppery flavour.

TWINSPUR
Diascia mollis

Twinspur has a long flowering period and the blooms can be seen from spring to autumn in South Africa.

Found in the wild mountains of South Africa, close to the sea, as well as in grasslands and forests, this is also a popular garden plant grown around the world for its tender flowers. Diascias have an important relationship with an oil-collecting bee native to temperate areas of Southern Africa, the *Rediviva*. The female bees are attracted by a window inside the base of the flower made up of a thin layer of cells, as they seek the fatty oil which they use to feed their larvae, produced by the glands found in the spurs. Pollination is carried out as pollen is deposited on the bee's body while she collects the oil.

COMMON LUNGWORT
Pulmonaria officinalis

Beginning this month, common lungwort will begin to develop buds which will soon burst into flower.

Used since the Middle Ages to treat coughs and diseases of the chest, the name 'officinalis' is given to refer to these supposed medicinal properties. Christian doctors at the time believed the plant was intended to cure the lungs because of the doctrine of signatures: a concept whereby God indicated which plant would cure which part of the body due to a resemblance. It was believed that the mottled leaves were reminiscent of a diseased lung. The plant is widespread in Europe, growing in a range of areas from lowland forests to mountains.

WINTER ACONITE
Eranthis hyemalis

Growing winter aconite alongside snowdrops provides a great floral display in the winter months.

From the buttercup family, this flower spreads out along woodland floors, forming a carpet of yellow flowers. Native to France, Italy and the Balkans, it is also popular in gardens, flowering early in the year and growing happily under shrubs and trees. With sweeping landscapes coming into fashion in 18th-century Britain, this flower became very fashionable. That the plant is also resistant to deer and rodents due to its toxicity, is probably another reason for its popularity.

COMMON CAMELLIA
Camellia japonica

Hand-coloured copperplate stipple engraving by Oudet after a botanical illustration by Johann Jakob Jung from Lorenzo Berlese's *Iconographie du genre Camellia, ou description et figures des camellia les plus beaux et les plus rares* (1841).

Thought to have arrived in England having been mistaken for the tea plant *Camellia sinensis*, this plant became of interest due to its beautiful blooms. It is believed that the plant, the first of any camellia grown in the UK, was cultivated in England by Lord Robert James Petre (1713–42) flowering in his glasshouse in Thorndon Hall, Essex. At the time, it was a very rare and expensive plant, but by the 19th century it was understood that the plant could actually be easily grown outside, and became incredibly popular.

COMMON HAZEL
Corylus avellana

From mid-February, the elegant male catkins will develop and hang from tree branches in clusters.

Hazel trees bursting into flower are an attractive sight to behold in early spring. Every hazel tree will produce both male and female flowers; however, the female ones are tiny and bud-like, whereas the male flowers are the long catkins that dangle bright yellow from the branches. Each catkin is made up of 240 individual flowers and, when ripe, the slightest touch releases copious quantities of pollen into the air.

MEZEREON
Daphne mezereum

This compact shrub is popular due to its blooms that burst into colour this month.

Also known as February daphne due to its flowering time, this plant is native to large areas of Europe. During colonial times, it was introduced to North America, where it has since naturalized in parts of the USA and Canada. The plant sap causes skin irritation and was once used to create rosy cheeks by applying it to the face; however, this was found to be as a result of blood-vessel damage and is not an advisable practice.

ALMOND
Prunus amygdalus

Marking the arrival of spring, and seen before many other blossoms, the artist Vincent van Gogh was a particular fan of this flower. He painted it many times; for him it symbolized new life. Notably, he painted this as a gift for his newborn nephew, who was named after him. Upon hearing the news of the birth, he said: 'It does me, too, more good and gives me more pleasure than I could express in words.' Fans and art critics comment that his joy upon the birth can very much be seen in his brushstrokes.

Almond Blossoms by Vincent van Gogh (1890).

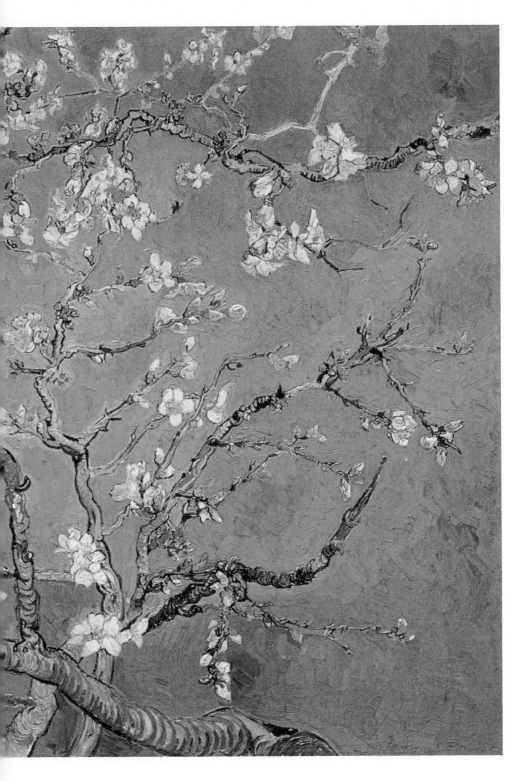

WHITE FORSYTHIA
Abeliophyllum distichum

The star-shaped flowers of this plant are borne on bare stems and have an almond scent; glossy green leaves emerge following the flowers, and these then turn purple in autumn. The plant is native to Korea and, while related to the common forsythia, the notable difference is that the flowers are white rather than yellow. Although popular in gardens, it is now rare to find in the wild in Korea due to over-collecting, and is classified as Endangered according to the International Union for the Conservation of Nature Red List.

BLEEDING HEART
Lamprocapnos spectabilis

Above: A suitably romantic gift on this day for Valentine's, this plant will reveal heart-shaped blooms just a couple of months later in April.

Left: Producing clusters of flowers before leaves appear, the white forsythia can be spotted in bloom this month and next.

This popular, shade-tolerant garden flower is native to Siberia, Japan, northern China and Korea. It was brought over from Asia to England and thence to North America in the 1800s and was enjoyed for its arching stems decorated with rows of heart-shaped flowers. The flowers are said to symbolize romantic love. For a more contemporary look, *Lamprocapnos spectabilis* 'Alba' has pure white flowers rather than pink. When the flower is upside down and the two outer petals are pulled apart, the shape of the flower resembles a lady, providing the plant with another common name, lady-in-a-bath.

QUAKING ASPEN
Populus tremula

This tree is grown for its shimmering foliage that ripples even in a slight breeze. Native to the cooler regions of Asia and Europe including the UK, the flowers appear as catkins, growing from mid- to late February, with male and female borne on separate trees. Following pollination, which is carried out by wind, the female catkins release fluffy, seed-bearing capsules during the summer months. The fluffiness of the capsules assists with their dispersal far from the parent tree.

EARLY BULBOUS IRIS
Iris reticulata

Native to Turkey, the Caucasus, northern Iraq and Iran, the dwarf flowers have a sweet fragrance. *Reticula* comes from the Latin meaning 'net-like', and the bulbs in this group are identifiable by the net that encloses each of the bulbs. Popular as a garden plant grown in temperate countries, the bulbs are hardy and will cope well with cold weather and come back each year.

Right: As the name suggests, this robust daffodil cultivar 'February Gold' can be found in flower this month.

Opposite top: Ahead of the fluffy seed heads making an appearance, the catkin flowers in early spring.

Opposite bottom: *Iris reticulata* 'Katharine Hodgkin' is a popular cultivar for late-winter delicate dwarf blooms.

DAFFODIL 'FEBRUARY GOLD'
Narcissus 'February Gold'

This daffodil is popular as an early-flowering variety. Its outer petals are swept back slightly, making it part of a group known as cyclamineus daffodils (the name referencing the similarity with the reflexed petals of cyclamen). This much-loved hybrid is reported to have been created in Holland in 1923 by the De Graaff Brothers by crossing *Narcissus cyclamineus* and *Narcissus pseudonarcissus*.

VIBURNUM
Viburnum × bodnantense 'Dawn'

Viburnum x bodnantense 'Dawn' flowers in abundance from November through to March.

This strongly fragranced plant was created at Bodnant Garden, now a National Trust property, which is open to the public in north Wales. The garden was established in 1874 by the scientist and politician Henry Davis Pochin who, along with his family, filled the garden with new plants being acquired by famous plant collectors of the time, including Ernest Wilson and George Forrest. In 1934, head gardener Charles Puddle crossed *Viburnum farreri* and *Viburnum grandiflorum* to create the hybrid and cultivar 'Dawn', which is notable for its attractive pink flowers and purple-pink anthers.

WALLFLOWER 'BOWLES MAUVE'
Erysimum 'Bowles's Mauve'

Wallflower 'Bowles Mauve' remains a firm favourite with gardeners, in part because it can bloom from February to October and attract pollinators such as the orange tip butterfly which will visit from spring.

Despite being possibly the most popular wallflower grown in gardens, the origin of this famous cultivar is not well understood. However, we do know for whom the plant was named: Edward Augustus Bowles (1865–1954) was one of England's greatest amateur gardeners and is also connected to current English royalty. He was the great-uncle of Andrew Parker Bowles, who was the first husband of Camilla, Duchess of Cornwall – now wife to Charles, Prince of Wales. Bowles developed an important garden at Myddelton House in Enfield (which is now open to the public), and encouraged gardeners by writing about plants and providing his own illustrations.

Right: Paperbush shown in a hand-coloured copperplate engraving by George Barclay after an illustration by Miss Sarah Drake from *Edwards' Botanical Register* (1847).

Opposite top: Grown in gardens for winter interest, the big clusters of flowers will start to appear this month.

Opposite bottom: White, delicate flowers of the plum tree appear this month, at the same time as the new leaves.

PAPERBUSH
Edgeworthia tomentosa

Also known as *Edgeworthia chrysantha*, this plant is native to China and the Himalayas, and is popular in gardens due to the richly perfumed, bright yellow flowers it produces early in the year, before the leaves appear. The common name refers to the Japanese use of the bark fibres for making handmade tissue known as *mitsumata* paper, which is very durable and even used to make bank notes.

EARLY STACHYURUS
Stachyurus praecox

Originally from Japan, the flowers of early stachyurus appear before the leaves in late winter to early spring. The flowers are borne on abundant, drooping, pendant flower spikes, making this an attractive winter garden plant. The genus name *Stachyurus* comes from the Greek *stachys* meaning 'ear of corn', and *oura* meaning 'tail' which relate to the appearance of the plant. The word *praecox* is from the Latin meaning 'very early' and is commonly found in the names of plants that flower early.

COMMON PLUM
Prunus domestica

From the end of this month, white blooms begin to appear on this popular and useful tree. Many cultivars are edible without having to be sweetened, while some have been bred for different tastes, sizes and garden use. Native to Southwest Asia, but having been grown in Europe for over 2,000 years, the flesh of the European-grown plum (*Prunus domestica*) is drier, with a much lower water content than the Japanese species, and is better for storing dried as prunes.

JAPANESE PINK PUSSY WILLOW
Salix gracilistyla 'Mount Aso'

Japanese pink pussy willow provides vibrant colour for gardens and landscapes in February.

This willow cultivar, of which the original species is native to Japan, Korea and China, is popular as an ornamental plant. The fuzzy pink catkins swell in mid- to late winter and provide colour when much else is dormant. This cultivar is thought to have been selected by a Japanese cut-flower grower as a particularly ornamental plant, which is used in floral arrangements, holding well in displays even without water.

CHINESE FRINGE FLOWER
Loropetalum chinense

Chinese fringe flowers come in white as well as pink and red varieties, blooming from February until April.

A little-known relative of the witch hazel, this plant is native to woodlands in China, Southeast Asia and Japan. It is also known as the fringe or strap flower, and may be more recognizable in the reddish leaf and pink flower cultivars that are more commonly grown in gardens of temperate countries, such as 'Fire Dance' or 'Ruby Snow'. The genus name *Loropetalum* comes from the Greek words *loron* meaning 'strap' and *petalon* meaning 'leaf' or 'petal' and refers to the fringe-like blooms.

SILK TASSEL BUSH
Garrya elliptica

The silk tassel bush makes for a popular garden plant, producing catkin flowers in January and February.

Native to North America, this evergreen shrub produces attractive tassels, making it popular with gardeners. The plant was collected in Oregon by one of the most famous Victorian plant collectors, Scotsman David Douglas, in 1828, and named after the deputy-governor of the Hudson's Bay Company which controlled the fur trade throughout North America. Coincidentally, the grey, flowering tassels of the plant are themselves almost furry in appearance. The popular cultivar 'James Roof' is grown for its particularly grey catkins. Once flowering has finished, the catkins remain on the plant for months, making it an attractive garden feature.

This illustration shows Siberian corydalis as the first plant on the left, in a chromolithograph by Henry Noel Humphreys after an illustration by Jane Loudon from *Mrs. Jane Loudon's Ladies Flower Garden of Ornamental Perennials* (1849).

SIBERIAN CORYDALIS
Corydalis nobilis

First introduced to Europe by Carl Linnaeus, the Swedish botanist who established the formal binomial nomenclature naming system still used for plants today, this flower is native to areas spanning Central Asia to southwestern Siberia and Mongolia. Supposedly Linnaeus had asked his friend, the explorer Erik Laxmann to bring back seeds of the bleeding heart plant (*Lamprocapnos spectabilis*), but was sent these instead from a Siberian mountaintop. The flowers resemble small snapdragons, while the seeds have a fatty body attached to them, which can be carried and eaten by ants, helping with dispersal, without harming the seed itself.

WINTER WINDFLOWER
Anemone blanda

Native to Greece and the eastern Mediterranean, this plant arrived in the UK in the early 1890s and was very popular with the likes of garden designers such as William Robinson and Gertrude Jekyll. The flowers lend themselves well to mixed herbaceous borders, as well as 'wild' gardens that were so in fashion around that time. In *The English Flower Garden* (1883) Robinson proclaimed that *Anemone blanda* was, 'deserving to be cultivated in every garden', extolling the virtues of its colour, hardiness, dwarf size and early flowering.

JAPANESE ANDROMEDA
Pieris japonica

The showy buds that were produced by this plant in midwinter will just be starting to open up, providing delight in the garden when many plants are still dormant. Native to parts of China and Japan, it is also known as the lily-of-the-valley shrub due to its many white, dangling clusters of flowers that last two or three weeks. Growing naturally in mountain thickets, its place within the heather family (Ericaceae) can be seen in the shape of the flowers.

Above: A bridge in Hirosaki Park provides a wonderful viewing spot from which to admire the cherry blossom.

Opposite top: from the end of this month, through March and April, anemone flowers burst open on sunny days.

Opposite bottom: Japanese andromeda cultivar 'Dorothy Wyckoff'.

CHERRY
Prunus × yedoensis 'Somei-yoshino'

In Japan, the blossoms are known as *sakura* and hold great importance. The time at which they flower not only marks the start of spring, but also a new school year for the children. They are celebrated as representing a time of renewal, and the fact that the blossoms only last two weeks is a reminder of how beautiful but brief life is, and to appreciate it. Social gatherings are held both during the day and night, where friends and family gather to eat, drink and look at the blossoms together from under the trees.

TULIP 'SEMPER AUGUSTUS'
Tulipa 'Semper Augustus'

Tulips with yellow and pink roses in a glass vase, oil on canvas, by Jan Philip van Thielen (1618–1667).

Originally found as wildflowers in Central Asia, tulips were prized in Persia and were a symbol of the Ottoman Empire in the 15th century. However, it was in Holland in the 17th century where Tulip Mania really exploded, with the Dutch so keen on this plant that vast sums were paid for the bulbs. This tulip, known as 'Semper Augustus' was the most valuable, and, as with other streaked tulips, it was a virus (tulip breaking virus) that caused this streaked colour but also weakened the plant. At the time, a single bulb of this tulip held the same value as a desirable house.

WILD DAFFODIL
Narcissus pseudonarcissus

Recognised as a welcoming sign of spring, wild daffodils make an appearance this month in grassy verges, damp woodlands and meadows.

I wandered lonely as a cloud
That floats on high o'er vales and hills,
When all at once I saw a crowd,
A host, of golden daffodils;
Beside the lake, beneath the trees,
Fluttering and dancing in the breeze.

'I WANDERED LONELY AS A CLOUD', WILLIAM WORDSWORTH (1804)

Considered a classic of English Romantic poetry, the first stanza of this famous poem captures the spiritual interaction of humans with their environment, an important aspect of the artistic movement known as Romanticism (c. 1800–50). The poem was inspired when Wordsworth and his sister Dorothy happened upon a wide belt of wild daffodils near their home in the Lake District, which you can still visit for inspiration today.

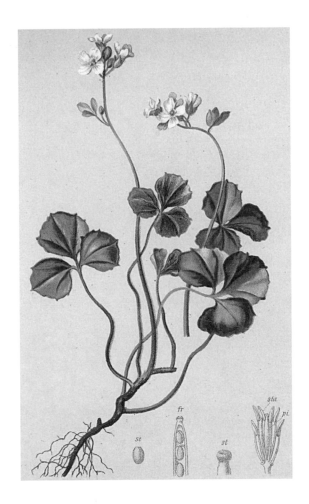

THREE-LEAVED CUCKOO FLOWER
Cardamine trifolia

Botanical illustration displaying the plant and floral parts of the three-leaved cuckoo flower from *Atlas der Alpenflora* (1882).

Native to many parts of the world, this plant produces small, cress-like flowers, which start to bloom as early as this month. *Trifolia* refers to the three leaflets that make up the foliage. As a shade- and somewhat drought-tolerant plant, it is used as groundcover in tricky woodland areas for ornamental purposes. The plant is a type of bittercress and produces pretty, ruffled white blooms.

GLORY-OF-THE-SNOW
Scilla forbesii

Making an appearance this month and next, glory-of-the-snow is one of the earliest bulbs to flower.

Native to the mountainous regions of Turkey, Crete and Cyprus, the common name of this plant comes from the fact that it often blooms sufficiently early that its flowers can be seen poking up through fallen snow. Each plant can have up to 12 star-shaped flowers, creating clusters of blooms. It is often grown *en masse* in gardens on lawns, or in rock gardens and woodland borders. The plants spread by creating bulbils on the side of the mother plant's bulb as it matures.

JUDAS TREE
Cercis siliquastrum

The bare stems of the Judas tree burst into blooms in March. The popular cultivar 'Bodnant' is grown here as a focus point in a garden amongst small shrubs.

Native to woodlands in the Mediterranean, this is often grown as an ornamental tree due to the purplish-pink flowers that bloom profusely. The flowers appear before the foliage, growing on the bare branches, making for a spectacular display. The common name comes from the belief that Judas Iscariot hanged himself from the tree after he betrayed Jesus; however, it is also thought to derive from an alternative name, Arbor Judea, as the tree was commonly cultivated around Jerusalem.

WEEPING FORSYTHIA
Forsythia suspensa

One parent of the popular garden plant *Forsythia × intermedia, F. suspensa* is native to China. It was spotted in a Japanese garden at the end of the 18th century and from there it started to be sold in Holland and England. Traditionally the fruits were used to treat inflammation in China, Japan and Korea. Known as the Easter tree, it is believed to symbolize anticipation, which is appropriate as the bare twigs in winter become full of cheery yellow flowers in spring, followed by the leaves.

Right: Billie
Holiday,
photographed by
William Gottlieb
c.1936, displaying
gardenia flowers
in her hair.

Left: The bright
flowers of
Weeping forsythia
bloom from
early spring.

COMMON GARDENIA
Gardenia jasminoides

Native to southern China and Japan, gardenias are prized for their strong fragrance. They produce a heady floral scent and are an expensive ingredient for perfume, with synthetic versions often used, or else a mix of other white flowers such as jasmine and orange blossom to re-create the scent. The performer Billie Holiday first wore gardenias in her hair to hide a patch that had been burned with a curling iron, but liked the look so much that she started wearing them for every performance

MIMOSA
Acacia dealbata

Mimosa flowers placed around the head of Michelangelo's David in Florence as part of International Women's Day celebrations.

Native to southeastern Australia, the flowers of this tree are used by florists. It is given as a gift on International Women's Day, which is celebrated on 8 March in parts of Europe and America, but has its roots in Russia where in 1917, the women of St Petersburg demonstrated against Russia's involvement in the First World War and against the rationing of food. After the end of the Second World War, in 1946, the women of Italy began to celebrate the date with bouquets of mimosas being offered to teachers, mothers, sisters and wives. The mimosa was chosen as a symbol of strength, sensibility and sensitivity.

WALLFLOWER
Erysimum bicolor

Flowering almost year-round, this is the wild relative of the more commonly known garden flower cultivar 'Bowles Mauve'.

This member of the cabbage family can be found growing on old walls or cliffs, and the flower is regarded as a symbol of fidelity. It is a good source of nectar for pollinators as it has a long flowering period. It was likely introduced from the Mediterranean but has been cultivated in gardens for so long that nobody is quite sure of its origin. A person is referred to as a 'wallflower' if they are shy and remain on the sidelines of social activity, much like the flower growing on a wall.

SPECIES IRIS
Iris damascena

Temple statues of King Thutmose III in Karnak, near Luxor, Egypt. Photograph from the archive of the Leo Baeck Institute, c.1930s.

King Thutmose III of Egypt (c.1479–26 BCE) was an avid fan of irises and perhaps one of the first to use them as garden plants. Unexpected treasures he found while conquering Syria were the many different irises growing there. He introduced them back home and they became very popular, symbolizing for the Egyptians the essence and renewal of life. They believed that the three petals of the flowers stood for faith, wisdom and valour.

STRIPED SQUILL
Puschkinia scilloides

A harbinger of spring, the striped squill starts to make a n appearance in March.

Native to western Asia and the Caucasus, this plant is named after the Russian chemist and plant collector, Count Apollos Apollosovich Mussin-Pushkin, who came across it in the early 1800s while on a botanical expedition, and introduced it to European gardens. The flower has a spicy fragrance and is admired for the blue stripes along each of its petals. It is a dwarf bulb that is popularly used at the front of garden borders or in woodland settings.

Right: Dog's tooth violet flowers pop up in woodlands and gardens this month and next.

Opposite top: Bird's nest banksia blooms in Australia's mid-summer to autumn months.

Opposite bottom: Sweet violet flowers are edible and used to decorate cakes as well as used for fragrance.

DOG'S TOOTH VIOLET
Erythronium dens-canis

It is the shape of the bulb – long and white – that lends itself to the name 'dog's tooth' rather than the appearance of the flower. This species is the only one of the genus native to central and southern Europe. Each plant produces a single flower at the beginning of spring, in white, pink or lilac. The leaves are edible and have been eaten in salads, while the bulb of different varieties has been processed to create a starch to make pasta and noodles in various parts of the world, including Japan.

BIRD'S NEST BANKSIA
Banksia baxteri

This plant is found only in Western Australia, growing in sandy dunes where nutrient levels are poor. It is popular with florists due to the plant's dramatic, lemon-yellow oval flower spike, said to look like a bird's nest. Banksias are adapted to survive poor-nutrient conditions as they produce cluster roots, also known as proteoid roots. The roots form as a thick mat just below the leaf litter and are able to enhance the nutrient uptake that they receive, possibly by chemically modifying the soil to make the nutrients more soluble.

SWEET VIOLET
Viola odorata

Less common than its almost identical but unscented relative the dog violet (*Viola riviniana*), sweet violet can be found growing on woodland edges. While usually blue in colour, it also occurs in white or lilac. Legend has it that you can only smell sweet violets once as they steal your sense of smell. There is some truth to this, as they do contain beta-ionone, a chemical known to temporarily shut off smell receptors.

AIR PLANT
Tillandsia ionantha

The air plant can be tied to branches for display. It is shown here with another popular type of air plant known as Spanish moss.

A relative of the pineapple, this plant is native to Central America and Mexico, and is believed to be naturalized in parts of Florida. A popular house plant, it thrives in bathrooms and other humid, well-lit rooms. Air plants absorb moisture and nutrients from the air and rain, producing colourful flowers. In the wild the plants attach themselves to trees, benefitting from the moist environment. The parent plant will slowly start to die after flowering but it produces small offset plants that continue to grow.

PUSSY WILLOW
Salix caprea

The yellow pollen of the male pussy willow flowers can be seen on mature catkins.

Also known as goat willow, this willow's male and female flowers grow on separate trees, meaning each tree has a distinct gender. These flowers have no petals and are known as catkins. It is the male flowers, which start out as oval, grey fuzzy pads, that are thought to resemble cats' paws, from which the common name is derived. As they mature, the pollen develops and turns them yellow, with the female catkins becoming longer and green. The plants are wind-pollinated and the female catkins later develop into woolly seeds. The branches of the male pussy willow are those favoured by florists for ornamental use.

WHITE CLOVER
Trifolium repens

A common plant native to grassy areas of the UK, Europe and Central Asia, its leaves form the symbol known as the shamrock. While each leaf usually comprises leaves of three leaflets, occasionally you may find one with four, which is considered lucky. Growing clover in lawns is a big help to pollinators, as its flowers are highly attractive to bumble and honey bees, producing an abundance of nectar, as well as being a food source for the common blue butterfly.

HELIOTROPE
Heliotropium arborescens

Above:
Heliotrope
produces masses
of purple flowers
in March.

Left: Illustration
of Saint Patrick
holding a
shamrock in a
stained-glass
window of
a church in
Kilkenny, Ireland.

A popular garden plant around the world, the heliotrope is native to Bolivia, Colombia and Peru, with sweetly fragrant flowers produced in large clusters. The name comes from the Greek *helios* meaning 'sun' and *tropos* meaning 'turn', as it was mistakenly believed that, like the sunflower, the flowers of the heliotrope followed the sun. In Australia the plant has become invasive, causing problems on grazing land as it is toxic to sheep, cattle and horses.

CRIMSON FLAG LILY
Hesperantha coccinea

In the wild,
crimson flag
flowers in South
Africa's late
summer and
autumn months.

Belonging to the iris family of flowers, this plant is native to southern Africa and Zimbabwe, but also a popular garden plant in Europe. It has bright red, star-shaped flowers, with the name *Hesperantha* meaning 'evening flower'. Large butterflies such as the Table Mountain beauty and certain flies pollinate the flower. It is found growing near water or on the edge of marshes, which is why it is also known as the river lily.

This highly-scented flower can be found in bloom both in the wild and in European gardens.

COMMON HYACINTH
Hyacinthus orientalis

Native to central and southern Turkey, northwestern Syria and Lebanon, hyacinths are grown in many parts of the world, both in gardens and indoors, where they are forced to flower early with artificially warm temperatures. According to folklore, it is said that sniffing fresh hyacinths can help with depression and grief, even preventing nightmares. According to Greek legend, the flower grew out of the blood of Hyacinthus who died during a game of discus with the god Apollo, although there is some debate about which flower is actually referenced in that story.

POLYANTHUS 'GOLD LACED'
Primula 'Gold Laced'

Above: Each flower of primula 'Gold Laced' has a golden centre with deep contrasting petals.

Opposite top: The evergreen star jasmine puts on an abundant display of fragrant flowers in spring.

Opposite bottom: Chinese quince produces exquisite blood-red blooms.

Polyanthus plants have thick stems that carry multiple flowers, as opposed to primroses, which have individual flowers on each stem. They are a natural hybrid of primroses (*Primula vulgaris*), thought to be produced from crossing with native cowslips (*P. veris*), the resulting flower known as a false oxlip (*P.* × *polyantha*). The term 'polyanthus' first appeared in the 17th century as the flowers became popular for their showy appearance. A particular favourite with the Victorians was the bred flower *Primula* 'Gold Laced' with golden and black petals.

STAR JASMINE
Trachelospermum jasminoides

Star jasmine is not actually a species of jasmine (*Jasminum*) at all, but instead resembles jasmine and is much easier to grow. It can withstand much cooler temperatures but has very similar fragrant flowers and in many areas does not die back in winter. Native to China and Japan, star jasmine is a source of oil used in perfumes. It is also known as 'trader's compass' due to a saying from Uzbekistan that as long as travelling traders were of good character, the plant would point them in the correct direction that they needed to go.

CHINESE QUINCE
Chaenomeles speciosa

Native to eastern Asia, this deciduous shrub produces showy, scarlet-red flowers on mature growth before leaves appear, making for a particularly spectacular display. For this reason, it is sometimes trained against walls to show off the flowers to best effect. After the blooms have faded, the plant produces yellowish-green fruit that are bitter when eaten fresh, but can be used in preserves, much like the fruit of the true quince.

COLTSFOOT
Tussilago farfara

Coltsfoot develops flowering buds and blooms before the leaves grow in early spring, which is why it is sometimes called son-before-father.

The bright yellow, daisy-like flowers of this plant emerge in spring, before the leaves appear. The leaves themselves are attractive, with a silver-white tint on the undersides. Coltsfoot was historically used as a cough remedy, and is sometimes called coughwort. However, it has actually been found to have a toxic effect on the human liver. Natural historians, as far back as Pliny the Elder in the 1st century CE, recommended smoking coltsfoot for its beneficial effects, and while this was later disproved, the plant went on to be used as a substitute for tobacco.

Botanical
hand-coloured
engraving on
copper of *Citrus
x aurantium* from
Mordant de
Launay's *Herbier
General
de l'Amateur,*
Audot (1820).

SEVILLE ORANGE
Citrus × aurantium

The highly fragrant flowers of the Seville orange are associated with good fortune. Orange-blossom water is made using the petals, and is popular in French and Middle Eastern cuisines, especially in desserts and baked goods. A smooth honey, known as orange-blossom honey, is created by moving beehives close to groves of orange trees; the resulting honey is mild and sweet, with a high vitamin C content as an additional benefit.

BILBERRY
Vaccinium myrtillus

Bilberry can
be found growing
in woodlands and
heaths, such as
shown here
in Finland.

Awild relative of the blueberries we buy in grocery stores, the bilberry
has higher levels of anthocyanins, which are beneficial for circulation.
This is due to the fact they have not been bred for improved flavour
and size at the expense of their nutritional content. Bilberry is native to
continental northern Europe, the British Isles and into northern Asia and
western North America. The flowers are small, white and tubular, followed
by the edible fruit. It is believed that soldiers in the Second World War
ate bilberries to improve their night vision, and the fruit is still eaten in
accordance with this belief by some pilots today.

QUINCE
Cydonia oblonga

Detail from
*The Story of
Oenone and Paris*
by Francesco di
Giorgio Martini
(c.1460), depicts
Paris awarding
the golden apple
(quince) to
Aphrodite.

This small tree is native to western Asia, and in order to flower requires temperatures below 7°C (45°F) for at least a fortnight. The flowers are produced from solitary buds, coloured pink and white. It is believed that the golden 'apple' given to Aphrodite by Paris in Greek mythology was a quince fruit, and today quince is still traditionally baked into Greek wedding cakes due to its reputation as an aphrodisiac.

CROWN IMPERIAL
Fritillaria imperialis

The exotic blooms of the crown imperial flowers shoot up from underground bulbs in spring for a spectacular display.

Native to southwestern Asia and across to the Himalayas, these flowers are bold and attractive, and are used around the world as the centrepiece for floral displays. The bright orange flowers hang from the top of the plant, and a glance inside them reveals large nectar drops held by the six nectaries borne by each flower. In Iranian folklore, the plant is associated with grief, either for the death of loved ones or religious or mythological deities, denoted by the drooping flower heads and the nectar 'tears'.

PASQUEFLOWER
Pulsatilla vulgaris

Known as
the Easter
flower, these
UK-native
blooms are
now rare but
can still be
spotted in chalk
and limestone
grasslands.

Flowering around Easter time, this flower is also known as the 'anemone of Passiontide'. It is native to Europe and southwestern Asia, with the hairy flowers emerging from the ground. It is now a rare plant in the UK due to changes to grassland management, particularly with the reduction in grazing-animal populations since the 18th century. In the UK the plant is associated with the country's violent past: legend has it that the flowers sprang up in places soaked by the blood of Romans or Danes in battle as they are found mainly on boundary banks.

HEART-LEAF BERGENIA
Bergenia crassifolia

From late winter and into spring, the flowers of this popular garden plant can be spotted.

Also known as pig squeak, due to the noise produced by rubbing the leaf between finger and thumb, and elephant's ears due to its shape, this plant is native to central Asia but is grown as a common garden plant in temperate countries. The leathery leaves are large and produced in rosettes that are heart-shaped at the base; the previous species name *cordifolia* is Latin for 'with heart-shaped leaves'.

POPPY ANEMONE
Anemone coronaria

The showy flowers of poppy anemone will start to make an appearance this month.

Originally from northern Africa, southern Europe and western Asia, the poppy anemone is now grown as a popular garden flower around the world. Since 2013 it has been the national flower of Israel, and each year there is a month-long festival to celebrate the blooming of the flowers. The Hebrew name, *kalanit*, refers to a bride, suggesting that the flower is as beautiful as a bride on her wedding day.

AUDREY II
Little Shop of Horrors (1986)

Audrey II
shown with
some of the other
cast members.

Named after a florist working in New York, USA, this rare Venus fly trap feeds on the blood of Homo sapiens and is capable of consuming entire bodies. Given favourable conditions, this plant will thrive and grow in size unusually fast. Native and endemic to outer space, the plant is vulnerable to electrocution, which can be used to control and destroy particularly challenging specimens.

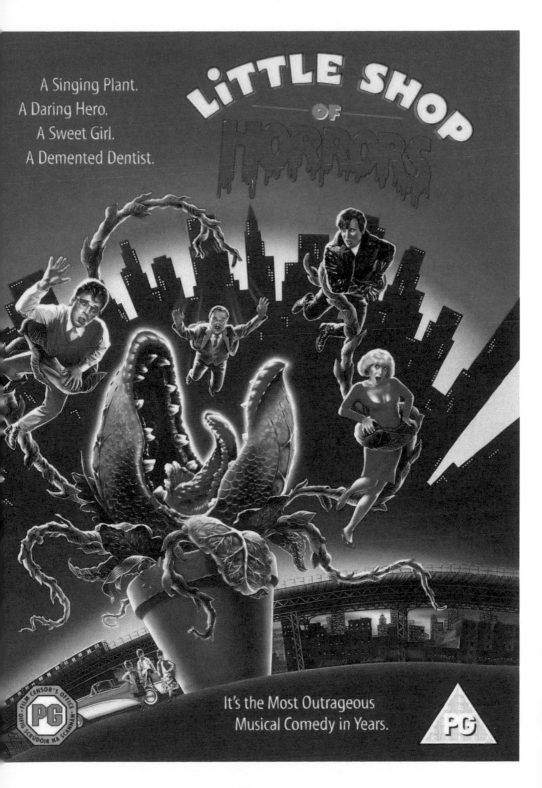

A Singing Plant.
A Daring Hero.
A Sweet Girl.
A Demented Dentist.

LITTLE SHOP
OF
HORRORS

It's the Most Outrageous
Musical Comedy in Years.

SNAKE'S HEAD FRITILLARY
Fritillaria meleagris

Although now a rare sight, from this month to next, snake's head fritillary will appear in damp meadows.

Native to Europe and western Asia, in the UK the plant is mainly found in England. The chequered purple, pink or white, bell-shaped flowers of these distinctive plants were once commonly found by rivers such as the Thames, where children would pick them to sell at flower markets. However, the loss of so many wildflower meadows means that snake's head fritillaries are now a much rarer sight. The common name refers both to the somewhat snake-like appearance of the nodding flowers and the scale-like pattern of the flowers.

A Polynesian girl wears leis made of frangipani as well as blooms over her left ear.

FRANGIPANI
Plumeria rubra

Originally from Central America and Mexico, the flowers have a delicious scent, comparable to that of a spicy rose or jasmine, and are mostly fragrant at night to attract pollinating moths. On Pacific islands such as Hawaii, where the flowers are known as *melia*, they are used for making garlands known as 'leis'. In modern Polynesian culture, wearing the flower over the left ear for women can mean they are in a relationship, while over the right ear suggests they are single.

Above: In mild springs bluebells start to bloom from the beginning of the month.

Opposite top: Delicate blooms of the foxglove tree burst from the branches this month, creating a spectacular display.

Opposite bottom: This small flowering dogwood tree will be graced with blooms this month and next.

ENGLISH BLUEBELL

Hyacinthoides non-scripta

F ound in ancient woodlands, blue carpets of blooms appear before many other wildflowers, creating a delicate floral scent, especially on sunny days. The introduction of the Spanish bluebell to England by the Victorians now threatens the native flower as it is much more vigorous in growth and even hybridizes with it. While Spanish bluebells have flowers all around the stem, English bluebells can be identified by their drooping stems with flowers on just one side. Almost half of the world population of bluebells is found in the UK.

FOXGLOVE TREE
Paulownia tomentosa

Also known as the empress tree, a whole tree full of large foxglove-like flowers is a sight to behold. It used to be customary in Japan to plant a foxglove tree when a girl was born, as it was fast-growing and large enough in time to be cut down and made into wooden items for her dowry. The woman would also be presented with a chest made of the wood carved from her tree on her wedding day.

FLOWERING DOGWOOD
Cornus florida

This small, deciduous tree is native to eastern North America. Considered one of the most beautiful of the native American trees, it is the state tree of Missouri and Virginia. The flowers are, in fact, tiny and yellowish green in colour; however, the petal-like white bracts surrounding each flower give it the appearance of large blooms. That the tree is also low-branching makes the flowers easy to enjoy as a spring display.

COMMON RHODODENDRON

Rhododendron ponticum

From the Ancient Greek meaning 'rose tree', there are many species of rhododendron, with this one native to parts of southwest Europe and western Asia. Its ability to produce many seeds and the roots to quickly produce new shoots, means it has become invasive in many parts of western Europe and New Zealand. It has also been discovered that the nectar of this flower (and some other species) contains grayanotoxins, which when ingested by bees can produce what is known as 'mad honey', which can be toxic to humans.

Common rhododendron quickly spreads, as here around the Torridon Hills, in the Northwest Highlands of Scotland.

APPLE
Malus domestica

Above: Different varieties of apple will bloom at different times between February and May; many will now be in full bloom.

Opposite: Dwarf crown imperial flower will start flowering now and into May, putting on a display of charming flowers.

For the most part, apple trees cannot self-pollinate and therefore must be grown with another apple tree of a different variety nearby in order to produce fruit. Most apple blossoms will start off pink and fade to white over time on the tree. Thought to symbolize love, peace and fertility, the flowers would be used by the Celts to decorate their rooms to encourage romance. Apple blossom is also believed to represent life continuing after winter, a long life, as well as life after death.

DWARF CROWN IMPERIAL
Fritillaria raddeana

This bulb is regarded as a smaller and more delicate version of the crown imperial (see page 96), both of which belong to the lily family, and it is shorter, with more delicate, paler flowers. It is found in rocky areas of Iran, Turkmenistan and the western Himalayas, but is also grown in gardens as an ornamental plant. The scent is described as disagreeable, but the attractive cluster of blooms on top of each thick, single stem make it desirable for garden displays.

SUMMER SNOWFLAKE
Leucojum aestivum

Known as summer snowflakes, these flowers follow on from snowdrops which will by now have gone over.

Found across most of Europe, *Leucojum* is derived from the Greek for 'white violet', but despite its common name, it blooms in mid-spring rather than summer. Often compared to a snowdrop, it is much larger in size, reaching a height of 60cm (24in) tall and is sometimes planted to follow on once snowdrops have gone over. All the fused six petals and sepals (known as tepals) of the fragrant flowers are of equal size, unlike snowdrops, which have three inner shorter tepals forming the inner 'cup' and three longer outer tepals.

SWEET ALYSSUM
Lobularia maritima

Sweet alyssum will continue to bloom, and more profusely into the summer months, but you may see it start its long flowering season this month.

Native to islands of south-eastern Europe, this plant has naturalized around many temperate parts of the world. It is commonly found on beaches, but also grows in fields, on walls, slopes and waste ground. Flowers have a honey-like fragrance and are produced over a long season, blooming so profusely that it can be hard to see the foliage. The flowers are very attractive to bees and butterflies, as well as tiny but beneficial insects that are able to access the nectaries inside the small flowers.

SNOWDROP TREE
Halesia carolina

This month
the snowdrop
tree will be
in full bloom,
and continue
flowering into
May, fed on here
by an orchard
oriole bird.

Native to the lower mountain slopes of the southeastern USA, the many bell-shaped flowers of this plant bloom just before or as the leaves appear. Also known as the Carolina silverbell, the name refers to South or North Carolina where it can be seen in the wild. The flowers are a good source of nectar for a wide variety of bees and birds, including the orchard oriole (pictured). The flowers are followed by green, four-winged fruit once they go over.

GREEN ALKANET
Pentaglottis sempervirens

From now and for the next two months, green alkanet will display forget-me-not-like blooms.

Often mistaken for borage during autumn and winter when not in flower, this invasive plant is native to Western Europe, and produces small blue flowers that are edible and make attractive decorations for salads and cakes. The common name is thought to be derived from the Arabic word for henna, suggesting that the plant was used as a cheaper substitute for dye. Bumblebees in particular enjoy the nectar, but once established, this bristly plant can be very hard to get rid of, with deep tap roots that are difficult to remove completely, making it less than ideal as a garden plant.

JAPANESE MAPLE
Acer palmatum

From the middle of spring, the tiny flowers of the Japanese maple will make their appearance among the new leaves.

Native to Japan, China and Korea, this maple produces tiny, reddish-purple flowers in mid-spring. Only visible when viewed close up, the flowers are umbel-shaped, with both male and female produced on the same tree. Following the flowers are winged seedpods, sometimes called helicopter seeds – which describes the way they spiral through the air as they fall. This mechanism allows them to travel further as they fall, giving them a better chance to develop into a new tree. It is believed that in the 15th century, Leonardo da Vinci came up with a design for what would much later become a helicopter, based on his study of maple seeds and their spinning mechanisms.

Capuliferae.

Quercus pedunculata Ehrh.

ENGLISH OAK
Quercus robur

This chromo-lithograph, produced after a botanical illustration by Walther Muller from *Köhler's Medicinal Plants* (1887), depicts the leaves as well as flowering and fruiting parts of an English oak.

The oak tree produces separate male and female flowers, both of which can be spotted in spring. The male flowers are greenish-yellow coloured, drooping catkins found growing in clusters; while the female flowers are small and pinkish, surrounded by a cluster of scales, and it is from these that the acorns later develop. The flowers are wind-pollinated, but self-incompatible, meaning it takes pollen from another tree to be accepted by the female flower. The well-known phrase, 'Mighty oaks from little acorns grow', is believed to be a proverb from as far back as the 14th century.

HANDKERCHIEF TREE
Davidia involucrata

Above: By April the handkerchief tree will have burst into bloom, displaying its dove-like inflorescences.

Opposite: From now until June, the buds of the sycamore tree will open into flower, with the seeds ripening from September to October.

Also known as the ghost tree or dove tree, this plant is native to woodlands in southwestern China. While the flowers themselves are red, they are covered with large, oval-shaped white bracts, which resemble handkerchiefs hanging from the plant in rows. It only takes a slight breeze to make the bracts flutter like birds on the branches, hence the name 'dove tree'. The species was introduced to Europe and North America from China in 1904 and became a popular ornamental tree.

Sycamore
Acer pseudoplatanus

Native to central and southern Europe and western Asia, this type of maple produces panicles of greenish-yellow flower clusters. The petals of each flower are small, and only on closer inspection is it possible to distinguish the male and female flowers. The female flowers are located at the top of the panicle (to discourage pollen falling from the male flowers and the plant self-fertilizing); below them are the male flowers, with the sterile flowers at the base.

EASTER LILY
Lilium longiflorum

Believed to symbolize purity, rebirth, hope and new beginnings,
the Easter lily is associated with the resurrection of Christ. It is
believed that some of the flowers were found growing in the Garden of
Gethsemane, where Jesus went to pray on the night before his crucifixion.
In many Christian churches, Easter lilies are used for decoration at this
time of year, celebrating renewal. Paintings such as *The Annunciation* by
John William Waterhouse (1914), also depict the Angel Gabriel presenting
Mary with lilies to announce that she would give birth to Jesus.

COMMON BAMBOO
Bambusa vulgaris

Golden bamboo is fast growing with attractive foliage and golden stems, as the name suggests.

Native to Indochina and tropical Asia, bamboo is one of the world's largest and most widely cultivated plants. Flowering is uncommon and the plant will live for many years before producing flowers, an event that is followed by the death of the plant. It is believed that the pollen has low viability, which is why fruit is not produced afterwards. The plant survives by producing offshoots that take around seven years to mature into clumps.

Right: From
this month
perfumed blooms
start to appear,
sometimes
followed by the
unusual sausage-
shaped fruit.

Opposite:
Starting in April
and flowering
into summer,
columbine
produces many
blooms which,
once gone over,
self-seed and
spread quickly
and readily.

SAUSAGE VINE
Holboellia coriacea

This evergreen vine is popular as an unusual plant to grow in temperate gardens around the world, but originally came from temperate east Asia. It is fast-growing and the scent is described as a mix of jasmine and melon. Its fruit is purple in colour and sausage-shaped, similar to an elongated plum, leading to its common name. The fruit is edible, and its roots and stems have been used in traditional Chinese medicine.

COLUMBINE
Aquilegia vulgaris

Native to Europe and North America, this is a quintessential cottage-garden plant found in woodland areas and damp grasslands. The name columbine comes from the Latin word for dove, *columba*, as the flowers are said to resemble a group of doves clustered together. Different species of *Aquilegia* are pollinated by different insects or birds, and the varying lengths of the spurs at the back of the flower where the nectar is held reflect this. In the UK, the garden bumblebee with its long tongue feeds from the plant; while in parts of the USA, such as California, the spurs of the Sierra columbine are up to 5cm (2in) long, and are pollinated by the much larger hawk moth.

SERVICEBERRY
Amelanchier lamarckii

Before the dark berries develop in July, the serviceberry tree puts on a display of big, white flowers.

From the rose family, this tree is also known as snowy mespil or juneberry. Native to North America, it is now also a popular tree to grow in Europe. It produces showy and fragrant, white, star-shaped flowers, which are followed in June by edible, dark purple berries, resembling blueberries in colour and taste. These berries are used in cooking for jams and fruit pies and are also popular with birds.

A woman strips the flowers off stems, in order to flavour wine in *Cowslip Wine* by Arthur Hopkins (1909).

COWSLIP
Primula veris

This plant used to be a common sight in traditional hay meadows and ancient woodland, but with the decline of those habitats it is a much rarer sight. It is thought that the name of this pretty flower has a less attractive origin: the plant was often found growing among manure in cow pastures, and the name may derive from the old English for cow dung. More appealingly, the flowers have been used in England to flavour wine, and the leaves eaten as salad in Spain.

LILY OF THE VALLEY
Convallaria majalis

This plant can be found in woodlands throughout Europe, away from coastal areas, and bears sweetly scented, bell-shaped flowers. It is believed to have been the designer Christian Dior's favourite flower and his company created a fragrance simulating the scent in 1956. Lily of the valley can be a popular, if expensive, choice for wedding bouquets and was featured at the wedding of Prince William and the Duchess of Cambridge in 2011. It is said to signify the return of happiness.

For only three weeks, lily of the valley blooms will flower, starting either at the very end of this month or at the beginning of the next.

STAR OF BETHLEHEM
Ornithogalum umbellatum

From now and into May, drifts of star of Bethlehem flowers can be spotted in gardens, meadows and woodlands.

These star-shaped flowers lend themselves perfectly to their common name, and in the wild these bulbs are widespread, from Africa to Europe and the Middle East. According to one legend, after guiding the Wise Men to the place of Jesus's birth, God thought the star too beautiful to extinguish and caused it to fall to Earth, shattering into pieces and creating the flowers we know today. The flower is used in religious ceremonies to signify innocence, hope and forgiveness.

BLUE WILD INDIGO
Baptisia australis

The seedpods of blue wild indigo are used to make dye. These follow on from the pea-shaped flowers which are in bloom from April to June.

Native to parts of central and eastern North America, this plant can be found growing along streams, in open meadows and at the edge of woodlands. It is enjoyed as a garden plant due to the intensity of its deep blue flowers. People of the Cherokee tribes, and later European settlers in America, traditionally used the seedpods as a blue dye. Legend has it that it can be planted around the home for protection. The plant is toxic to eat and care should be taken as the young shoots can be mistaken for asparagus.

GRAPE HYACINTH
Muscari armeniacum

Above: This month and next, grape hyacinth will be in flower, capable of growing in a range of challenging situations.

Opposite: If you are unable to visit the Philippines, many larger botanic gardens with glasshouses around the world will likely house this exotic plant.

Native to the woodlands and meadows of western Asia and southeastern Europe, these flowers are also known as bluebells or bluebonnets in North America. The name *Muscari* comes from the Greek meaning 'musk' and refers to the scent of some of the species. The name 'grape hyacinth' comes from the resemblance of the flowers to small bunches of grapes and to the form of hyacinths.

JADE VINE
Strongylodon macrobotrys

From the tropical forests of the Philippines, this climbing plant is prized for its showy turquoise flowers. In the wild, bats will hang upside down from the flowers and drink the nectar inside. The claw-shaped flowers hang in grape-like clusters on pendants that can reach 3m (10ft) long. Much admired in botanical glasshouses around the world, in the Philippines where the flowers are enjoyed as a food source, the plant is endangered due to deforestation.

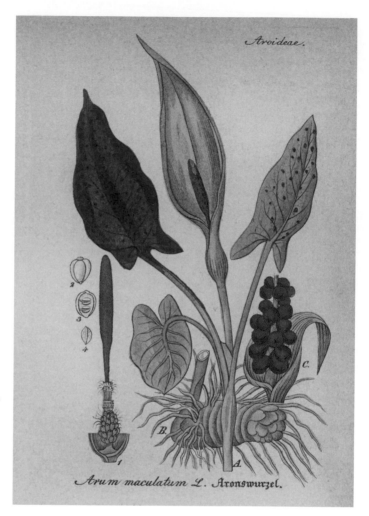

Hand-coloured copperplate engraving depicts the parts of the lords-and-ladies plant, from Dr Willibald Artus' *Handbook of all medicinal pharmaceutical plants* (1876).

LORDS-AND-LADIES
Arum maculatum

Native to much of Europe and northern Africa, this plant grows in wooded areas, its name referring to the way the flowering parts resemble both male and female reproductive organs in humans. It is unusual to find this common plant in flower; it is more commonly spotted in autumn, when the lower ring of flowers forms a cluster of bright red berries, decorating the woodland floor and providing food for birds.

MILLION BELLS
Calibrachoa parviflora

Million bells have a long flowering season, from this month through to the end of summer.

These trailing flowers are related to petunias and are also native to parts of South America and southern North America. The name million bells refers to the many flowers borne on each plant and *Calibrachoa* species attract hummingbirds where they occur. The flowers are smaller than those of petunias, but the plants tend to be hardier, making the cultivars bred from this plant popular in gardens.

MAYFLOWER
Epigaea repens

ound in North America, it is believed that the flower was named by the Pilgrim Fathers after their ship, as the plant was found in abundance where it made landfall at Plymouth Rock in Massachusetts. It was subsequently chosen to be the state flower. The plant itself is a small, creeping shrub with small five-lobed flowers, in either white or pink.

MOCK ORANGE
Philadelphus coronarius

Above: The
many flowers
of the mock
orange shrub
will bloom from
late spring into
early summer.

Left: A replica
of the Mayflower
ship displays a
representation
of a mayflower
painted on to
its stern.

Native to southern Europe, the common name is derived from the appearance of the flowers, which are similar to those of oranges and lemons. The scent of the flowers is also reminiscent of orange, but with a hint of jasmine, and is attractive to pollinators including bees and butterflies. The plant is grown for its scented blossom in gardens in temperate regions, providing the appearance of an orange tree where it is too cold to grow one.

AZALEA RHODODENDRON
Rhododendron indicum

The evergreen azalea rhododendron flowers this month and next, producing a flush of exotic colour.

Xiang shu my dear,
You make me think of home.
Of luxurious days spent pondering your beauty.
Of the scent of Spring that lingers around your roots.
DU FU (712–770 CE)

In Chinese culture, azaleas are celebrated as being a 'thinking of home bush' (*xian shu*), representing a type of mindfulness. Azaleas were made famous by the work of the Tang dynasty poet, Du Fu, who is considered by many critics to be the greatest Chinese poet. Du Fu influenced many writers , including William Shakespeare. Having been selectively bred for hundreds of years, there are now over 10,000 cultivars of this flower, a member of the Rhododendron family. Azaleas are widely planted throughout parts of southwest Europe, Asia and North America.

BLUEBLOSSOM
Ceanothus thyrsiflorus

Blueblossom
flowers for
around six weeks
from late spring,
and sometimes
even again
in October.

Found in Oregon and California in the USA, this plant is also known as the Californian lilac. An evergreen shrub, it is popular with pollinators, such as butterflies, birds and honeybees. The word *thyrsiflorus* comes from the Ancient Greek, roughly translated as 'with flowers arranged in the shape of a contracted panicle or wreath'. Alongside other *Ceanothus* species, this plant plays an important role in Californian eco-systems, as they are among the first plants to colonize sites burned in wildfires.

AMAZON MOONFLOWER
Selenicereus wittii

While the
Amazon
moonflower
will take a while
to develop each
flower bud, they
will each only
open fully for
a single night.

Found growing on tree trunks in the forests of the Amazon, this cactus produces tall, white flowers borne on flower tubes up to 27cm (10½in) long. The slender, pure white tepals intensely reflect ultraviolet light. Each flower only opens for a single night, typically beginning to open after sunset, and is fully open within two hours. Before the flowers are fully open, each gives off an intense fragrance, although this turns into an unpleasant odour as they continue to ripen. In the wild there are only two types of hawk-moth with long enough tongues to pollinate the moonflower as the nectar is stored at the base of the flower tube.

JAPANESE SNOWBELL
Styrax japonicus

In May and June, Japanese snowbells burst into flower; the cultivar 'Pink Chimes' is shown here.

Native to China, Japan and Korea, and found on forest edges, this tree produces showy and fragrant white flowers. The waxy blooms are formed in clusters and are bell-shaped; they are easily visible from below as the foliage has an upward posture. As the plant was first collected in Japan and introduced to the West in 1862, it is known as Japanese snowbell, although it is believed to have been long cultivated in the East before it was grown in botanic gardens of the UK and elsewhere.

GORSE
Ulex europaeus

Flowering from early spring to late summer, common gorse can be seen in towns as well as heaths and coastal grasslands.

One area this common plant is found is along the borders of fields: farmers would plant this thorny shrub to keep intruders out and livestock in. The tender shoots provide food for cattle and sheep, so farmers burn back the old growth, which then also provides nutrients for the soil. It is a very long-flowering plant, often in flower for most of the year, which is why there is the old saying, 'When the gorse is out of bloom, kissing is out of season'.

WHITE FRINGETREE
Chionanthus virginicus

The showy white blooms of the white fringetree make their appearance in May and June.

Also known as snowflower and old man's beard, this tree produces delicately fragrant, white flowers in panicles up to 20cm (8in) long. The trees are separately male or female, with the male specimens producing the showier flowers. The white fringetree is native to the lowlands and savannas of the USA, but can also cope with colder temperatures and so is grown further north. Native American peoples traditionally used the dried roots and bark to treat skin inflammations.

T. 2. Nº 13.

EUROPEAN HORSE CHESTNUT
Aesculus hippocastanum

The flowers with their showy stamens of the European horse chestnut as depicted in *Treaty of Trees and Shrubs* by Pierre-Joseph Redouté (c.1800–1830).

Native to the Balkan Peninsula, this plant was introduced to the UK from Turkey in the late 16th century. The white flowers of this tree appear with yellow guide spots; these turn red after they have been pollinated, which are less attractive to insects, thereby directing them to unpollinated flowers. Each tree can live for over 300 years, and they are famous for producing glossy, red-brown conkers, which fall during early autumn and are traditionally strung with string and used in conker fights by children. The first recorded game of conkers was on the Isle of Wight in 1848.

BLUE THROATWORT
Trachelium caeruleum

The highly scented flowers of blue throatwort bloom over a long season, typically starting in early summer.

Native to the Mediterranean, this plant produces large sprays of flowers that are attractive to bees. Both the common and botanical names refer to its use as a treatment for throat ailments, *trachelos* meaning 'neck' in Greek. It reseeds readily and is becoming more popular in gardens with an informal or cottage-garden feel. When using as cut flowers, stems should be picked when only a quarter of the flowers are open; they will then last up to two weeks in a vase.

WISTERIA
Wisteria sinensis

The long tresses of lilac flowers make this plant instantly recognizable. Often associated with old houses in Europe and North America, each plant can live for over a hundred years. It will train easily up support and produces an abundance of sweet-smelling, pea-like flowers in the spring. It is believed to symbolize love and a long life, making it an appropriate plant to grow up the walls of a home. While it has been popular in the West since John Reeves, Chief Inspector of Tea in Canton, southern China, first brought it over as a plant cutting to Britain in 1816, it had been cultivated as an impressive garden plant in Japan and China for a much longer period.

Wisteria flowers around April to June but may also have a second flush in August. Gardeners are mindful to not remove next year's buds when pruning the long stems after flowering for impressive displays.

BYZANTINE GLADIOLUS
Gladiolus communis subsp. *byzantinus*

Illustration of *Gladiolus communis* (left) shown alongside *Schizanthus pinnatus* (centre) from a Japanese hand-coloured woodblock print (c.1900).

Originally from southern Europe and northern Africa, this plant can be found growing wild in the UK, as a garden escapee. The name 'gladiolus' comes from the Greek word *gladius*, meaning 'sword' due to its shape. It is believed that gladiators in Rome would wear the corms of the plant around their necks in battle as a symbol of strength and integrity. It also symbolizes infatuation – as the sword-like shape represents the ability to pierce a heart – as well as remembrance, which is why it is associated with fortieth wedding anniversaries.

GUELDER ROSE
Viburnum opulus

Following flowering in June and July, red berries develop on the guelder rose and are used in food and drink such as to decorate traditional Ukrainian bread.

Guelder rose is a deciduous shrub that grows alongside old hedgerows, in woodland edges and beside rivers in the UK, Europe, northern Africa and central Asia. The larger flowers around the outside of the flower clusters are sterile, functioning purely to attract the attention of pollinators to the smaller, fertile flowers within. The name 'guelder rose' comes from the Dutch province of Gelderland, where the snowball tree (*Viburnum opulus* 'Roseum'), a popular cultivar, is meant to have originated. References to the flower can be found throughout Ukrainian folklore, including poetry, embroidery and other visual arts. The red berries are thought to symbolize blood and home.

COMMON LABURNUM
Laburnum anagyroides

From the mountains of central and southern Europe, the laburnum is a popular tree belonging to the pea family and is grown for its huge ornamental appeal. Known also as the golden rain plant, in spring long garlands of yellow flowers hang from the branches. First introduced to Britain in the 1560s, it is an important feature of many gardens from the Royal Botanic Gardens, Kew, to Claude Monet's garden in Giverny in northern France. Gardeners put in a lot of work pruning and caring for this plant, whose flowers only bloom for 2–3 weeks, but create a breathtaking sight.

From mid- to late May and into June, the glorious golden flowers of the common laburnum burst open along pendulous racemes.

COMMON LILAC
Syringa vulgaris

A young faun playing pipes, reminiscent of the Greek god Pan from the watercolour *Idyll* by Marià Fortuny (1868).

Grown for its heavily and sweet-scented flowers, lilac was brought over to northern Europe at the end of the 16th century from Ottoman gardens in the Balkans. According to Ancient Greek mythology, Pan, the god of forests, fell in love with a nymph named Syringa, who turned herself into a plant in order to hide from him. When he found the shrub, he made panpipes from its hollow stems.

PEONY
Paeonia lactiflora

Peony flowers
are at their peak
in the month of
May, producing
many glorious
blooms.

A traditional floral symbol of China, where it is known as the 'king of flowers', the ancient city of Luoyang in central China has a reputation as the centre for the cultivation of these plants, with many shows and exhibitions held there annually. Peonies are popular in gardens as well as in bouquets, despite only flowering for a few weeks each year. Their impressive, ruffled petals, large flower heads and soft, fresh scent make them a classic garden choice, believed to symbolize good luck, honour and love.

YELLOW CORYDALIS
Pseudofumaria lutea

Yellow corydalis have a very long flowering period, starting in the middle of spring and continuing to produce flowers until the first frost of the year.

Also known as *Corydalis lutea*, the Greek word *korydalis* means 'lark', and it was used in reference to the resemblance of the spurs on the flowers to those on the feet of the bird. With its fern-like leaves, this plant is native to the foothills of the Alps in Italy and Switzerland, but can be found widespread in the temperate Northern Hemisphere, either in gardens or woodlands or cracks in walls. It was brought over to the UK in the 16th century, and has many local common names including fingers and thumbs in Dorset, gypsy fern in Sussex and poppers in Wiltshire.

PERSIAN BUTTERCUP
Ranunculus asiaticus

Persian buttercups flower well this month and next, with continual blooms appearing.

This plant blooms for around six weeks, with each flower in a vase lasting up to seven days, making double-flowered cultivars (which have many petals) of this plant particularly popular in ornamental gardens for cut flowers. The cultivars have large, rose-like flowers and are grown from corms (similar to bulbs), which are best planted in autumn. Cultivated for centuries, ranunculus originate from northern Africa, southwestern Asia and southern Europe.

FRINGE CUPS
Tellima grandiflora

Native to moist forests in western North America, this plant has since become naturalized in places such as the UK and Ireland, believed to have escaped from gardens into the wild. The flowers can be fragrant and fade to a light red colour as they age. It is believed that the Skagit people of North America traditionally used the plant for medicinal purposes, such as to cure a loss of appetite. They would crush the plant and use it to make a liquid infusion.

LADY'S SMOCK
Cardamine pratensis

Left: From May to July fringe cups produce flowers on tall spikes, turning pink as they mature.

Above: Lady's smock will be in full bloom this month and next, producing many delicate pinkish flowers.

This is also known as the cuckooflower, as it is said to flower around the same time of year as the first call of the cuckoo, a summer visitor to the UK. The plant is believed to be sacred to fairies, and it is therefore regarded as unlucky to bring it indoors or use it to make garlands. It can be found throughout the UK, most of Europe and western Asia; it is also naturalized in parts of North America, probably as a garden escapee. The name is also believed to refer to the 'cuckoo spit' sometimes found on the plant, which is laid by an insect called a froghopper, rather than a cuckoo.

ELDERFLOWER
Sambucus nigra

Once the
flowering stems
of elderflower are
collected, blooms
are picked off
to steep and
produce a cordial.

Elderflower cordial was popular with the Victorians but can actually be traced back as far as the Roman era. The honey-scented blooms of the elder, which are high in vitamin C, are collected fresh when young and before it rains, before the pollen inside has been washed away. The flowerheads are steeped in a sugar solution with lemon juice, or similar, used to help preserve the drink. The mixture is then strained and diluted with water, tonic or gin.

HAWTHORN
Crataegus monogyna

Hawthorn is also known as May blossom as it predictably blooms this month.

Highly scented and growing in flat clusters, these flowers are white or sometimes pinkish. The plant is known as the fairy tree in Celtic mythology, as it was believed that fairies live under it, symbolizing love and protection. It is one of the most sacred trees and is still revered in Ireland today. In Ireland in 1990, work on the M18 motorway from Limerick to Galway was halted as the planned route threatened a fairy tree. A campaign launched by folklorist Eddie Lenihan led to the motorway being rerouted, and the road finally opened after a ten-year delay.

ROWAN

Sorbus aucuparia

The flowers, leaves and fruit of the rowan tree are depicted in this hand-coloured botanical illustration from *Spirit of the Woods* by Rebecca Hey (1837).

Native to western Asia and Europe, the rowan is regarded by some as a magical tree, believed to symbolize courage and wisdom, and providing protection from evil. In Neo-Druidism, it is known as the 'portal tree' and considered to be a threshold between this world and others. It is also believed that the tiny, five-pointed star opposite the stalk on each berry represents the pentagram, an ancient protective symbol.

WILD STRAWBERRY
Fragaria vesca

Between April and July the wild strawberry produces flowers which quickly develop into sweet, small fruit.

A member of the rose family, this plant produces edible fruit, which is much smaller than the fruit grown for commercial production, but sweeter and with a hint of vanilla. In medieval times, the Christian church decided that strawberries were the fruit of the Virgin Mary, with the case being made that not only were the leaves divided into three lobes, representing the Holy Trinity, but that also the five petals of the flowers represented Christ's five wounds and the red fruit, his blood.

SICILIAN HONEY GARLIC
Allium siculum

Sicilian honey garlic flowers in late spring and early summer, producing clusters of delicate blooms.

Native to Turkey, southern France, Italy and surrounding areas, this flower grows in damp, shady woods and produces attractive clusters of graceful, bell-shaped flowers. It is grown as an ornamental plant in gardens due to its showy flowers and twisted foliage, but is also used for culinary purposes. The leaves are used in spice mixes and seasoning in Bulgaria. In a similar way to its relative, the onion, it releases eye-watering chemicals when chopped or crushed.

DOG ROSE
Rosa canina

16th-century English stained glass with a dog rose in the centre depicting a Yorkshire family's heraldry.

In hedgerows and woodland edges, this thorny climber weaves between other shrubs to use them as support. It is a common symbol in English medieval heraldry, and in Germany it was believed that fairies would use the flowers to make themselves invisible. Once the flowers have gone over, the rose hips are consumed in the form of syrups to promote good health as they contain high levels of vitamin C.

RUSSIAN OLIVE
Elaeagnus angustifolia

The small yellow blooms with their sweet scent are produced in late May and June.

Native to Asia and neighbouring areas, the Russian olive is also now widely planted in North America. The common name is derived from its similarity to the olive tree, despite not being at all closely related. In traditional Persian spring celebrations, known as Nowruz, it is used as a table setting as it is believed to represent love. In Iran, the fruits are dried and the powder is mixed with milk to treat arthritis and joint pain.

TULIP TREE
Liriodendron tulipifera

The magnificent flowers of the tulip tree truly look like blooms that one expects to find in a flower bed, usually at their best from May and into June.

Native to eastern North America, the tulip tree produces large, cup-shaped flowers that resemble tulips. It is a popular tree for providing shade in gardens but it needs plenty of space as it can grow to more than 35m (115ft) tall and can live for hundreds of years. A member of the magnolia family, it is the state tree of Indiana, Kentucky and Tennessee. The flowers are popular with bees and the wood is used for timber, being lightweight and easy to work with.

COMMON CAMAS
Camassia quamash

Native to western North America, the deep blue flowers of the common camas appear in late spring, largely in prairies and marshes. It is grown ornamentally around the world, naturalized in grassy areas, but has also been a food source for the native people of the USA and Canada. Once the flowers have gone over, the bulbs are harvested, then boiled or roasted. When boiled, a syrup is produced. When roasted it is like a sweet potato, but with a sweeter taste; excessive consumption leads to flatulence.

FREESIA
Freesia refracta

The freesias we grow and buy today are bred from wild freesias, such as *Freesia refracta*, which has buttery yellow petals. Native to South Africa, it is used in perfumes and bouquets of cut flowers because of its attractive fresh scent. In the northern hemisphere, freesias bloom at the beginning of summer; gardeners who remember to plant their bulbs the previous autumn are treated to the earliest blooms.

ROSEMARY
Salvia rosmarinus

Found growing naturally throughout the Mediterranean region, Rosemary has been a plant sacred to many cultures. It was believed that the blue flowers were white, until the Virgin Mary hung washing over the plants during their flight to Egypt to keep the infant Jesus safe from King Herod, and her blue dress stained the plant. It is grown in monastery gardens and is believed to have many positive attributes, including protecting against evil. Recent scientific study has found that rosemary oil helps aid memory, something that has been believed for centuries, as documented by Shakespeare in *Hamlet* when Ophelia says, 'There's rosemary, that's for remembrance'.

Opposite top: While common camas may be spotted as early as April, it will be creating a show of colour now, producing flowers in clusters.

Opposite below: Signalling the beginning of summer, freesias can last up to three weeks in a vase making them a popular choice for decorating the home as well as growing in the garden.

Left: Ophelia depicted holding rosemary in the painting *Ophelia and Laertes* (1879) by William Gorman Wills.

Rosier à cent-feuilles, foliacé.
J.J. Redouté
Langlois

ROSE OF A HUNDRED PETALS
Rosa × centifolia

The rose of a hundred petals is illustrated in this hand-coloured stipple engraving *Centifolia Rose* by Redouté (1835).

With a scent revered above all others, this rose is the perfumer's choice. Created through plant breeding by the Dutch in the 17th century, it is now famously grown between the mountains and sea in the small town of Grasse, France and sold to some of the finest perfume makers in the world. An important component in Chanel No.5, the rose's scent is described as particularly complex with a dewy freshness. The extract from this flower is worth more than its weight in gold.

HYDRANGEA
Hydrangea macrophylla

Hydrangeas flower profusely from mid- to late summer on the plant's previous year's growth.

This flower is native to Japan and Korea, and while it can come in a range of colours from purple to red and pink, the most celebrated of them is the blue. It is the acidity in the soil that affects the colour of the blooms, with a higher level of acidity producing bluer blooms. Where soil is insufficiently acidic, hydrangeas can be grown in pots of ericaceous soil to manipulate the colour. In Japan, the flower is historically linked to gratitude and apology. It is believed that a Japanese emperor gave hydrangeas to a maiden he was in love with, to apologize for neglecting her.

GERMAN CHAMOMILE
Matricaria chamomilla

Even without the small daisy-like chamomile blooms, the plant will release a scent when underfoot, making it a pleasant alternative to a typical grass lawn.

Chamomile is one of the most ancient herbs used medicinally; of all the species it is German chamomile that is mostly used to treat pain and promote sleep. Flowers are picked and dried, with two teaspoons of dried flowers used per cup of tea, steeped for 10–15 minutes. Chamomile can be grown as an alternative to, or mixed with, grass for a lawn. As the plant is crushed underfoot, it releases the slightly apple-like fragrance, producing a relaxing aroma.

MEADOW BUTTERCUP
Ranunculus repens

Children play with buttercups, holding the flower under each other's chins to determine who likes eating butter.

Children play a game with this flower where they hold it underneath the chin of a friend, and the amount of yellow reflected there reveals how fond of butter they are. Physicists researching the reason for this unusual colour reflection have revealed that a carotenoid pigment present in the flower absorbs blue and green light, meaning that mainly only yellow is reflected back, appearing on skin. The reflection is amplified by the smooth surface of the petals, which is also good at reflecting ultraviolet light, resulting in an increase in the visibility and attractiveness of the flower to pollinators such as bees.

YARROW
Achillea millefolium

Dying Achilles, a statue created in Berlin in 1884, now a centrepiece of the Achillion Palace in Corfu.

Since the times of Ancient Greece, this plant has been used to treat wounds, and modern research has revealed that leaf extracts do in fact contain anti-inflammatory and antioxidant properties. The Latin name of the plant *Achillea* is thought to come from the Greek mythology that the warrior Achilles would use yarrow to heal soldiers fighting on behalf of the city of Troy. Achilles himself was bathed as a baby in the magical water of the river Styx to make him invincible. However, as he was held by his heel while being immersed in the water, this part of him was unwashed and left vulnerable, ultimately leading to his death.

CORNFLOWER
Centaurea cyanus

Flowering from June to September, cornflowers are a popular garden plant with many beautiful blue blooms.

It is believed that when Queen Louise of Prussia was being pursued by Napoleon's forces in 1806, she hid her children in a field of cornflowers, which were growing there as weeds. The story goes that she kept them quiet and hidden by weaving wreaths for them to wear made from the flowers. Today, intensive farming in the UK means that wild cornflowers, which can grow up to 1m (3ft) high, have been almost completely wiped out, and are now a priority species for conservation.

FOXGLOVE
Digitalis purpurea

This is one of the most potent plants used medicinally. Small doses are used in traditional medicine to treat heart issues, calm fevers, ease pain and other ailments. However, it can also be toxic, sometimes due to it being misidentified as comfrey, and this can result in severe poisoning through ingestion. Legend has it that not only do fairies wear the flowers as hats, but they gift them to foxes to help them sneak silently into chicken coops at night.

LADY'S MANTLE
Alchemilla mollis

Above: Raindrops held on the leaves of lady's mantle, were believed to be the purest form of water by alchemists.

Left: Foxgloves flower from June to September, producing tall spikes of colourful blooms.

When it rains, the dense hairs on the leaves of this plant catch and hold droplets. The name *Alchemilla* comes from the belief of alchemists that the water found as droplets on the leaves were the purest form of water and these would be used when attempting to turn base metal into gold. The common name 'lady's mantle' is believed to come from the soft and therefore perceived femininity of the leaves.

EYEBRIGHT
Euphrasia officinalis

Eyebright starts flowering around June, creating small and dainty blooms, each only 5–10mm (¼–½in) in size.

The common name of this plant refers to the historical user of the plant in the treatment of eye infections. Nicholas Culpeper, the famous English botanist (1616–54) claimed that the plant could bring clarity of vision. Parts of the leaf, stem and small parts of the flower would be prepared as a tea or made into a warm compress. Many species are found in alpine areas and the flowers themselves can be said to have a 'bright eye', as the centre of the flower is yellow.

HIMALAYAN BLUE POPPY
Meconopsis betonicifolia

Himalayan blue poppies can be found in flower this month, growing well in damp ground and with dappled light.

Blue is one of the rarest colours found in flowers, and few are bluer than this. Greatly admired and much sought after, this striking flower is considered quite tricky to grow in gardens as it needs cool, moist and shady conditions to do well. But it does create a spectacular display in woodland and alpine areas in spring and early summer, with the paper-thin, sky-blue petals sat delicately upon upright stems. When seeds arrived from the Himalayas in 1924, collected by plant hunter Frank Kingdon-Ward, they were considered a sensation and still are today.

Angelica illustrated in a chromo-lithograph after a botanical illustration by Walther Müller in *Koehler's Medicinal Plants* (1887).

ANGELICA
Angelica archangelica

The name of this herb is believed to come from the dream of 14th-century botanist and medic Mattheus Sylvaticus in which the Archangel Michael appeared and told him that it could be used to treat victims of the bubonic plague. It was also thought that growing the plant or having it in the home provided protection from witchcraft. In fact, the plant has a history of being used for over 4,000 years in many parts of the world for traditional medicinal purposes for many ailments, including tuberculosis.

VALERIAN
Valeriana officinalis

Valerian was said
to be used by the
Pied Piper who is
depicted here in
an illustration by
H.J. Ford (1890).

This plant is used to promote sleep as it is believed to have a sedative effect on humans. It is taken as a tea or in capsule form, with the extract from the root having an effect similar to the drug Valium, but without any of the adverse effects. In contrast to its calming action on humans, it excites cats and rodents in a way similar to catnip. In the myth of the Pied Piper of Hamlin it is believed the protagonist used valerian to bait the rats and remove them from the city.

SORREL
Rumex acetosa

Sorrel develops spikes of small red and yellow flowers in summer, which can be used as well as the leaves in salad.

Also known as spinach dock, this deep-rooted plant can be found in grasslands throughout Europe. More recently it has also been introduced to countries like Australia and North America, growing well in poor soil. Around the world it is used as a food; the young leaves have a tangy taste due to the high content of oxalic acid, which can be poisonous in large quantities. The leaves are rich in vitamins A and C and historically have been used as a cure for scurvy. Both the leaves and flowers can be added to salads and used as a garnish.

SOLOMON'S SEAL
Polygonatum multiflorum

Solomon's seal flowers in May and June, with the blooms followed later by black berries.

The word *Polygonatum* comes from the Greek for 'many knees', which refers to the multiple joints along the underground stems (rhizomes) of this plant. Native across Europe to the Caucasus, it is valued as a garden plant due to its ability to grow well in shady conditions and provide elegant flowers on arching stems. The flowers are produced on the underside of the stems, perhaps explaining another common name, ladder-to-heaven.

GARDEN COSMOS
Cosmos bipinnatus

Above: Cosmos grow easily from seed and have a long flowering season, starting in early summer and sometimes lasting until frost.

Opposite: Lady's slipper orchid blooms from June into July, producing unusual-looking flowers growing on stems up from the ground.

Grown by Spanish priests in their mission gardens in Mexico, who named the flowers *Cosmos* due to their evenly distributed petals, thought to resemble the universe. The word *cosmos* comes from the Greek for 'harmony' and 'ordered universe'. The plants were eventually brought back to Madrid from Mexico, making their way to England in the late 18th century, supposedly by the wife of the English ambassador to Spain, and then on to the USA in the mid-1800s. Popular in gardens as they make good cut flowers, they also protect neighbouring plants by attracting aphids away from them.

LADY'S SLIPPER ORCHID
Cypripedium calceolus

Native to Europe and Asia, this plant is typically found in open woodland but is now extremely rare in the UK and by the late 20th century there was just a single native colony in England. The plant has since been grown in cultivation and reintroduced, with its location kept secret and monitored to protect it from theft. The name *Cypripedium* comes from a Greek phrase meaning 'Venus' foot' as it, is believed to resemble a small shoe.

WILD GARLIC
Allium ursinum

Wild garlic flowers are usually spotted in May and June, with the leaves tasting better picked before flowers develop.

Found in shady woods, this plant is an indicator of ancient woodland and is an important flower for bees and other insects. It is also a popular foraged plant as the leaves can be made into soup or a pesto, while both the leaves and flowers can be added to salads, providing a mild garlic taste. The best way to identify it is by its strong, pungent aroma, pointed leaves and flowers, each having a six-pointed star shape.

ROSE GRAPE
Medinilla magnifica

Rose grape
illustration by
Walter Hood
Fitch in Curtis's
*Botanical
Magazine* (1850).

Native to the humid mountains of the Philippines, but grown as a houseplant in cooler parts of the world, the rose grape produces large, attractive garlands of pink flowers that hang down in grape-like bunches. Found climbing up trees, where the blooms hang down, the flowers are followed by clusters of pink berries. This plant is an epiphyte, which means that it grows on trees or other plants, rather than on the ground, feeding off the rain and rotting vegetation around it.

FLOWERING SEA KALE
Crambe cordifolia

Flowering sea kale works well in garden settings such as growing behind lupins and foxgloves here at Hidcote garden in Gloucestershire, England.

Related to cabbage and other brassicas, this ornamental plant is native to the Caucasus. The second part of the name, *cordifolia* comes from the Latin, meaning 'heart-shaped' and refers to the leaves, which are edible and should be cooked for the tastiest result. The young flowering shoots can also be eaten, and are similar in taste and form to broccoli. When it flowers, the plant creates a huge spray of scented tiny, white flowers like a giant version of baby's breath (*Gypsophila paniculata*).

PEACE LILY
Spathiphyllum wallisii

Peace lilies will frequently flower indoors with sufficient heat, adequate watering and decent levels of light.

Popular as a houseplant around the world due its ability to grow in low light, this flower was spotted by plant collectors growing wild in Colombia. It produces flowers close together on a spadix in the middle of the inflorescence with a large white bract growing around them. It is often associated with purity, innocence and peace, as the name suggests, due to its lily-like white inflorescences.

BISHOP'S FLOWER
Ammi majus

Found in areas of southern Europe, north Africa, and parts of western and central Asia, this is a popular garden plant with pretty, lace-like umbels of white flowers – more delicate than the similar looking cow parsley. It is popular in floral arrangements and, when left unpruned, the seed heads are feasted on by birds such as finches. In Ancient Egypt it was used to treat skin conditions.

COMMON DOGWOOD
Cornus sanguinea

Above: Small white flowers are produced in clusters on common dogwood around June.

Left: Bishop's flower blooms from around late June to August, producing a froth of white inflorescences.

These creamy, four-petalled flowers appear in summer and are followed by colourful berries. Dogwood is native to most of Europe, as well as western Asia, and grown widely as an ornamental plant. The new stems are very bright red in winter and the tree is often cut short each year to encourage new growth for winter interest. The timber is so hard that it was used to build weight-bearing crucifixes and it is believed to have been used for the cross that Jesus was crucified on.

BEGONIA
Begonia gracilis

Begonias start flowering in June and continue doing so until the first frost, making them popular garden plants.

The first of many begonias to be written about in the West is what was to become known as *Begonia gracilis*, spotted in Mexico and written about in the 17th century. The many commercially available begonias sold today are hybrids made by crossing the wild species, of which there are over 1,500, found all over the world. The seeds of these flowers are some of the smallest, measuring just 0.2mm (1/100in).

HONEYWORT

Cerinthe major

Honeywort
flowers in June
and into July,
producing
clusters of bell-
shaped flowers.

Native to the open meadows and grassy plains of southern Italy, Greece and other parts of the Mediterranean, the name *Cerinthe* comes from the Greek word *keros* meaning 'wax' and *anthos* for 'flower'. It was thought that bees obtained wax from these flowers before the process of beeswax production was fully understood. The variety 'Purpurascens' (shown above) is common as a garden flower as it has particularly colourful bracts and purple tubular flowers.

PEACH-LEAVED BELLFLOWER
Campanula persicifolia

Peach-leaved
bellflower
depicted on
the San Marino
postage stamp
(1967).

A common flower of the Alps and other mountain ranges in Europe, the peach-leaved bellflower can be seen in woodland margins, broadleaf forests and meadows. It is commonly grown in gardens and regarded as an English cottage-garden classic. According to Greek legend, Cupid shot a shepherd boy in the hand in order to retrieve a magic mirror belonging to Venus that the boy had found. Once the mirror hit the ground, it shattered into many small pieces and bellflowers began to grow where each shard had fallen.

MEXICAN MARIGOLD
Tagetes erecta

In Mexico, marigolds are used to decorate altars filled with candles and photographs of relatives who have passed away, as seen in Disney's Pixar animation *Coco* (2017).

Native to Mexico and Central America, these flowers are used to decorate home altars during the annual, traditional Mexican Día de Muertos celebrations. It's believed that the scent of their bright orange blooms help attract souls to the altar.

This species is also among those used as a companion plant to help protect more valuable plants or crops from pests without the use of insecticides. A popular example of this technique is to grow marigolds between tomatoes, aubergines and chilli peppers, in order to keep whitefly away before populations of these pests can become established. This is believed to work as marigolds release the chemical limonene, which deters the pests.

ANISEED
Pimpinella anisum

Alongside liquorice, aniseed is used to make traditional sweets such as Liquorice Allsorts, first produced in Sheffield, England in 1899.

Native to southwest Asia and the eastern Mediterranean, this plant is used in teas and sweets around the world. Alcoholic drinks, including Greek ouzo, Italian sambuca and French absinthe, are flavoured with aniseed. It is grown in vegetable patches so that the smell can deter pests as well as attract insect predators of pests in an effort to avoid using pesticides.

PALE PITCHER PLANT
Sarracenia alata

The yellow flower of the pale pitcher plant can be seen nodding forward, behind the tall pitchers in front of it.

The modified leaves of pitcher plants are prey-trapping. Insects attracted to the plant slip down the pitchers and drown in a fluid at the base of the pitcher. The flowers of this species may be pale yellow, greenish or reddish and will bloom slightly ahead of the pitchers forming. As the flowers are held on long stems, this avoids potential pollinators being caught in the trap of the pitchers below. In addition, each flower head droops downwards, meaning the anthers containing pollen are also hanging down and can make easier contact with visiting pollinators.

CIGAR TREE
Catalpa speciosa

Above: The cigar tree produces showy white flowers with orange stripes and purple spots on the inside.

Opposite: Coral drops flower late into summer, producing more blooms as the season develops.

Native to the midwestern USA, this species is widely grown as an ornamental tree. The showy flowers have purple marks inside the petals and bright yellow stamens. Following the flowers are large, green, bean-like seedpods. The name *Catalpa* is derived from the name for the tree given by the Muscogee people in an area that is now within the state of Oklahoma, and *speciosa* means 'showy' and refers to the flamboyance of the flowers. A common parasite found on the tree, known as the catalpa moth caterpillar (*Ceratomia catalpae*), is highly sought-after by anglers as it is widely thought to be one of the best live baits for fishing.

CORAL DROPS
Bessera elegans

Originally from Mexico, this dainty bulb with elegant blooms was brought over to Europe by German plant hunter, Count Wilhelm Karwinsky in the early 1830s. The delicate red, lantern-like flowers that hang from the top of the plant make it popular in gardens. Looking inside the flowers reveals white markings creating patterns or stripes and a purple pistil (the female part ending in a stigma). Sometimes described as red snowdrops, the blooms hang from the top of long, wiry stems.

WILD SWEET PEA
Lathyrus odoratus

Sweet peas grown in gardens are bred from their wild relatives and will be in full flower this month.

This wildflower native to Crete, Italy and Sicily was first sent across Europe from Sicily in 1699 by the monk Franciscus Cupani. More significantly, in the late 19th century, it was developed into the decorative garden sweet peas that have become very popular and are so commonly grown today. Sweet peas are associated with comfort and a pleasant departure, believed to provide protection and good luck. The origin of the name comes from the Greek words for 'very passionate' and 'fragrant'.

CORPSE FLOWER
Rafflesia arnoldii

The bright red colour of the corpse flower helps it resemble flesh and attract flies to pollinate it. It grows up to 1m (3ft) across.

Boasting the largest flower in the world, this plant has no roots or leaves, does not photosynthesize, and instead feeds off other vegetation as a parasite by embedding itself into the plant to absorb nutrients. The fleshy flowers resemble and smell like rotting meat in order to attract and be pollinated by flies, and usually last just a week. Native to the rainforests of Sumatra and Borneo, this rare flower is one of three national flowers of Indonesia.

LOTUS
Nelumbo nucifera

Above: Lotus in flower at the Pura Taman Kemuda Saraswati temple in Indonesia.

Opposite top: Peruvian lilies are very popular flowers in the floristry industry.

Opposite below: Sea thrift shown on a British threepence coin.

Regarded as a symbol of purity and enlightenment in Buddhism and Hinduism, the lotus flower and leaves come up almost miraculously clean while growing even in the dirtiest of water. This has been found to be because the small bumps on the leaves causes them to repel liquids. The surface is superhydrophobic and in effect self-cleaning, as liquid on the leaves beads up and rolls off. This has become known as the 'lotus effect' and has been used to create technology for non-stick products such as pans and self-cleaning windows.

PERUVIAN LILY
Alstroemeria aurea

Resembling a miniature lily, these plants are native to South America, but are now grown in many places around the world and are very popular in floral arrangements. They can last up to two weeks in a vase, and as they have no scent can be enjoyed by those with certain allergies that make other flowers unsuitable. First brought over to Europe in the 18th century, they are believed to carry strong positive sentiments of devotion, friendship and prosperity.

SEA THRIFT
Armeria maritima

Found in coastal areas throughout the UK, Europe and North America, this plant is popular as a garden plant, in dry areas such as rock gardens. During the Second World War, the flower was used as an emblem on the threepenny coin as a reminder of the importance of spending money wisely. In the Outer Hebrides and in the Orkney islands, to the north of Scotland, it has been boiled with milk and used as a traditional remedy for tuberculosis, as well as consumed to cure hangovers.

COMMON SPOTTED ORCHID
Dactylorhiza fuchsii

This month, common spotted orchids may be seen on chalky soils, including some roadside verges.

This is the easiest orchid to spot in Europe and can even be found as far as eastern Asia; it has also been naturalized in Canada. It grows on woodland floors, roadside verges, old quarries and marshes, where it appears as a carpet of blooms. The name refers to the fact that the green leaves are spotted with purple. The scented flowers range from white through to purple and have distinctive pink spots and stripes on their three-lobed lips.

Giant water lily depicted in a 19th-century engraving. The flowers are enlarged for clarity, as it is actually the leaves which are surprisingly large, being able to reach up to 3m (nearly 10ft) in diameter.

GIANT WATER LILY
Victoria amazonica

This is the largest water lily species in the world and is native to tropical South America. In 1837, seed was collected in Guyana and brought over to the UK, with the plant subsequently being named after the reigning monarch Queen Victoria. Each fragrant flower only blooms for two nights: the first night it opens up with white petals as a female, then closes in the morning to reopen the following night pink as a male flower. This helps the plant to avoid self-pollination and is one of the world's natural wonders.

SEA KALE
Crambe maritima

Above: Sea kale is a summer-flowering plant, the name referring to its wild location, resemblance to kale and that it can be eaten.

Opposite: The bee orchid is usually only in flower for a couple of months each year, with blooms resembling bees, attracting them for pollination.

With honey-scented flowers and bluish leaves, this was a popular garden vegetable in the 18th century. It grows naturally on the shoreline, on shingle and rocky beaches. In common with other members of the cabbage family (Brassicaceae) each flower has four sepals. The family was previously known as *Cruciferae*, a reference to the number of petals representing the four parts of the crucifix. All parts of the plant, including the flowers, are edible.

BEE ORCHID
Ophrys apifera

This orchid provides an example of clever plant mimicry. Not only does the flower release the scent of a female bee, but the brown velvet lip of the flower with its yellow markings also resembles one, with the sepals looking like as wings. Male bees are fooled by this and will fly over and try to mate with it. When the male bee lands on the flower, pollen is transferred, and as they visit multiple flowers pollination occurs.

COMMON STOCK
Matthiola incana

Stock blooms
depicted in
a chromo-
lithograph by J.
de Pannemaeker
in Jean Linden's
*L'illustration
Horticole* (1885).

Native to coastal southern and western Europe, this plant has also naturalized in western parts of the Mediterranean. It is popular as a cut flower and grown in gardens, especially in England and the USA. The scent of the flowers is reminiscent of cloves, and is believed to represent happiness and contentment, making stock popular for weddings. The flowers can be air-dried when tied loosely and hung upside down to preserve the bouquets.

CALIFORNIA POPPY
Eschscholzia californica

Californian poppies are short-lived perennials that flower in masses during the summer months, and are grown as annuals in cooler climates.

The official state flower of California, chosen in 1903, as its golden colour is thought to represent the 'fields of gold' sought during the Gold Rush of the mid-19th century. It can be seen on some of the welcome signs along highways as you drive into the state. Native to the USA and Mexico, it is best enjoyed outside on a sunny day, as the petals close at night as well as on cloudy days; they also drop quickly once picked.

NARROW-LEAVED LUPIN
Lupinus angustifolius

Bred and grown mostly as an ornamental garden plant now, some species of lupin have a history of being used for food in the Andes, as far back as 6,000 years ago. The seeds were soaked then baked,

toasted or boiled to make both sweet and savoury dishes. The name lupin comes from the Latin for 'wolf' (*lupus*), as the Romans observed how vigorously the plants grew in fields. They thought that the plants were stealing ('wolfing up') nutrients, but in fact, as with other legumes, bacteria in the plant's roots extract nitrogen from the air and when the plant dies this is made available to other crops.

Lupins are beautiful garden flowers but are invasive in parts of the world such as New Zealand and the United States, crowding out other native plants.

CANDLE LARKSPUR
Delphinium elatum

Native to Europe as well as northern and central Asia, the name comes from *delphis*, the Ancient Greek for dolphin, due to the shape of the flowers. Larkspur species in North America are believed to have been commonly used by Native American people and European settlers to make blue dye, as well as to help with sleep and relaxation. The plant is toxic to both humans and animals, and has been used to repel scorpions, lice and other parasites.

Left: This month and last, candle larkspur will be in full bloom. Shown here is the cultivar 'Spindrift'.

Right: The Duchess of Cambridge holds a bouquet containing sweet Williams on her wedding day in 2011.

SWEET WILLIAM
Dianthus barbatus

This popular garden plant is native to southern Europe. There are many theories about where the name originates, including possibly being named after William Shakespeare, the 12th-century priest William of York or William the Conqueror. At the wedding of Prince William and the Duchess of Cambridge, sweet Williams were included in the bride's bouquet as a tribute to the bridegroom. For the Victorians, the plant symbolized gallantry. It is related to the carnation and the flowers are edible.

RED AMARANTH
Amaranthus cruentus

Red amaranth produces showy catkins this month. The cultivar 'Velvet Curtains' is shown here.

Red amaranthus is popular as a garden plant in the Northern Hemisphere with its long purple catkins produced in summer, but was first cultivated by the Aztec people in Central America, and processed in a manner similar to maize. It is still a popular snack in Mexico today, a symbol of indigenous culture. The Aztecs would make statues of their god using amaranth grains and honey and share it as part of their religious ceremonies.

FUCHSIA
Fuchsia triphylla

Fuchsia triphylla has long tubular flowers which are all in full bloom this month. Cultivar 'Thalia' is shown.

The first of the many fuchsias to be botanically described in the West, this plant was spotted on the Caribbean island of Hispaniola in 1896–97 by French monk and botanist Charles Plumier. Now popular garden plants, they are enjoyed for their months of back-to-back flowering in the summer, with some species able to survive winters in temperate regions like the UK. They are also known as 'lady's eardrop' as they resemble ornate earrings. The fruit that follows the flowers can be made into jam or jelly.

ACONITE
Aconitum napellus

Aconite
illustrated in a
hand-coloured
print by Mary
Ann Burnett,
from *Illustrations
of Useful Plants*
(c.1840).

Known also as monk's hood or wolfsbane, aconite is found mainly in western and central Europe, and is highly poisonous. While the common name monk's hood refers to the shape of the flower, the Latin name relates to the use of the plant juice on arrow tips to kill wolves. If ingested in large quantities, the toxic chemical aconitine found in the roots and tubers causes stomach problems and dizziness, followed by heart and respiratory failure, which can often be fatal.

CUP AND SAUCER VINE
Cobaea scandens

The cup and saucer vine will start flowering this month, producing large bell-shaped flowers.

Also known as Mexican ivy or cathedral bells due to the shape of the flowers, this vine is native to Mexico. It is grown widely in gardens in temperate regions as a tender plant, putting on a display of exotic-looking purple flowers in late summer. In the wild its flowers are pollinated by bats. The plant was named after a 17th-century missionary and writer, Bernabé Cobo, who wrote a history of the Inca people. It also caught the attention of Charles Darwin, who used it to study the mechanics of climbing plants.

WILD CARROT

Daucus carota

Portrait of Anne
of Denmark
on horseback
with Windsor
in the distance.
Engraving
by Simon de
Passe, (1616).

With lacy leaves and clusters of flowers borne in flat, dense umbels, this plant is also known as Queen Anne's lace. According to legend, the wife of King James I, Queen Anne, was challenged to create a lace flower by her friends. While doing so, she pricked her finger, resulting in a reddish stain in the centre of the flower, thus inspiring the common name for the plant. It is also similar in appearance to the poisonous hemlock plant, but without the purple mottling on the stems.

PURPLE BELL VINE
Rhodochiton atrosanguineus

Purple bell vine will flower from now until October, providing late-summer blooms.

Native to Mexico, this plant can be found growing in areas such as the margins of temperate rainforest and clearings in pine-oak cloud forest. It was illustrated and featured in the *Curtis's Botanical Magazine* in 1834 and went on to became popular with botanic gardens. Seed had been sent from Mexico to Munich in 1828 and thereafter it was shared around the world. It is now widely grown as an ornamental climber: where the climate is warm it is grown as a perennial; while in cooler regions it is grown for the summer each year from seed.

Above: Rosebay willowherb will usually be at peak flowering this month and next.

Opposite top: Despite Himalayan balsam being an invasive plant, it does benefit pollinators with its nectar-rich blooms.

Opposite bottom: The frilly blooms of the crape myrtle are out in full force this month, providing much colour.

ROSEBAY WILLOWHERB
Chamaenerion angustifolium

Native to temperate regions of Europe, Asia and North America, this flower, now regarded as a weed, was once only found in heaths, woodland clearings and mountains. However, it now grows in many gardens around the world. Related to evening primrose and fuchsias, each plant can produce a staggering 80,000 seeds, which makes it extremely proficient at spreading itself. It flourishes in disturbed and vacant areas, and was particularly successful in the UK following the upheaval of the Industrial Revolution and both World Wars. It became known as 'bombweed' believed to survive destruction, and as a consequence was no longer enjoyed in gardens, as it became associated with wartime.

HIMALAYAN BALSAM
Impatiens glandulifera

The invasive cousin of bizzy lizzies, this is a flower now found across much of the Northern Hemisphere. Native to the Himalayas, it has been introduced elsewhere and as it can withstand the cool, high altitudes of its homeland, it finds it easy to establish itself in places like the UK. This troublesome plant has colonized miles of riverbank and other damp places, competing with native plants. However, the flowers are nectar-rich, and as it flowers late into the year, it provides a source of food for bees later than many other plants.

CRAPE MYRTLE
Lagerstroemia indica

Native to China, Japan, Indochina, the Indian Subcontinent and Southeast Asia, this plant produces crinkled, crepe-like white, pink or purple flowers over a period of a few months. It enjoys hot weather, making it a popular ornamental plant in places like the southern states of the USA, where it can also somewhat withstand droughts. The bark and leaves have been used as laxatives in traditional medicine, and the seeds used to treat insomnia.

CHASTE TREE
Vitex agnus-castus

Illustration of the chaste tree depicting all parts of the plant, in the Byzantine manuscript known as the *Vienna Dioscurides* c.512 AD.

Native to the Mediterranean, northern Africa and western Asia, this plant produces showy, fragrant violet flowers. Used as a popular herbal treatment, studies have been carried out to research the plant's potential for the treatment of premenstrual syndrome, showing that it may be helpful in alleviating pain and beneficial in fertility treatment. This is understood to be because it contains dopaminergic compounds that stimulate the brain. It was also believed to quell sexual desire, relating to its name, while Plato described its aphrodisiac qualities, but as of yet both of these claims remain scientifically unproven.

GOOSENECK LOOSESTRIFE
Lysimachia clethroides

The inflorescences of gooseneck loosestrife are thought to resemble the necks and heads of a gaggle of geese.

The flower spikes of this plant are slender and arched, resembling the neck of a goose, and when seen as a group of flowers are impressive. The plant is originally from China, Japan and Indonesia. It is believed to evoke calm and tranquillity, which explains the use of the word 'loosestrife' in the name. According to ancient legend from Macedonia, King Lysimachus of Thrace used the plant to calm a wild ox.

GOLDEN-BEARD PENSTEMON
Penstemon barbatus

Golden-beard penstemon are fed from by hummingbirds, which will transfer pollen from one flower to another.

Native to the southern USA and Mexico, the flowers of this plant are very attractive to hummingbirds. It flowers in late summer, coinciding with the southern migration of the rufous hummingbird, which travels from Mexico to Alaska and back. This trip is the longest migration of any bird on Earth as measured in body lengths.

MOUNTAIN CAMELLIA
Stewartia ovata

Mountain camellia flowers on the tree branches around July, a less common time for other trees to be in bloom.

A member of the tea family, related to the camellia used to make the beverage, it is native to the southeastern USA, and is found on wooded stream banks and the base of cliffs from Virginia to Alabama. The attractive flowers are much like those of a camellia, with five white petals and orange anthers. The flowers are followed by woody seedpods that split open when ripe. In the autumn, the leaves turn orange and red, providing good seasonal colour.

SUNFLOWER
Helianthus annuus

Self Portrait,
Anthony van
Dyck (after 1633),
oil on canvas.

In this famous self-portrait, it is believed that the artist chose to paint himself with a sunflower as it expressed his devotion to King Charles I of England. The symbolism of the sunflower being that it turns its head throughout the day to follow the source of all life, the sun, just as the artist devotedly followed the king. In reality it is only young sunflowers that do this, doing so to receive maximum sunlight for photosynthesis. Once they are mature, they will stay facing east, to provide an attractively warm surface for bees.

Illustration depicting Prince Augustus Frederick, Duke of Sussex (1773–1843) wearing the robes of a Knight of the Order of the Thistle. The thistle emblem can be seen decorating his chest.

SPEAR THISTLE
Cirsium vulgare

This floral emblem of Scotland is likely to not be native to Scotland at all, but instead introduced there from mainland Europe sometime before the 16th century. However, today it is found abundantly in Scotland and has many associations with the country. Mary, Queen of Scots (1542–87) had the image of the plant incorporated into the Great Seal of Scotland as a symbol of longevity. The Order of the Thistle, a Scottish order of knights, was founded by James V in 1540 and adopted the thistle as part of their badge together with the motto, *Nemo me impune lacessit* ('No one provokes me with impunity').

SEA POISON TREE
Barringtonia asiatica

The sea poison tree produces many fine but showy stamens, and each flower is about 15cm (6in) wide.

Native to the mangroves of the islands of the Indian Ocean, western Pacific and surrounding areas, this plant produces very showy flowers with four white petals and many fine, pink-tipped stamens. Following the flowers are fruit known as box fruit due to their square shape. All parts of the plant are poisonous; the pulped fruit was used by indigenous peoples to help them catch fish and octopus.

COMMON HOUSELEEK
Sempervivum tectorum

Clusters of common houseleek flowers are produced on long stems protruding from the foliage.

Native to the mountains of southern Europe, this succulent plant is widely cultivated. It was believed by the Romans and others that growing it on buildings would protect against lightning strikes. In fact, it has been reported that the Holy Roman Emperor Charlemagne, who ruled much of Western Europe in the 8th and early 9th century, passed a law that houseleeks should be grown on the roof of every dwelling as a safety measure. While the leaves that form rosettes themselves resemble flowers, the plant does also produce blooms. The plant is monocarpic, meaning that once it flowers the plant dies; however, as it produces many offsets, it will appear to live on.

Above: The creamy blooms of meadowsweet flower prolifically in fields and meadows during the summer months. Shown here growing alongside rosebay willowherb (*Chamaenerion angustifolium*).

Opposite: Titan arum shown on its second day of flowering at the US Botanic Garden, Washington DC.

MEADOWSWEET
Filipendula ulmaria

Growing in damp meadows and riverbanks, this sweet-smelling flower produces frothy clusters of flowers on tall stems. The name derives from the Anglo-Saxon *meodu-swete* as it was used to flavour mead. In Yorkshire it has been called 'courtship and matrimony' as, once the sweet-scented flowers are crushed, the plant releases salicylic acid, which has an antiseptic smell, suggesting something a lot less enjoyable.

TITAN ARUM
Amorphophallus titanum

Found in the wild only in western Sumatra, Indonesia, this plant has the largest unbranched inflorescence in the world. Its flowering parts can reach a height of over 3m (10ft) and bear many small flowers. The name means 'giant, misshaped phallus', reflecting its unusual shape. It takes the plant up to 7 years before it has enough energy to flower, growing from an underground, bulb-like corm. Once in bloom, which will happen in an evening, it remains open for the night and typically only for the next 48 hours. While in flower, it heats up, with some parts reaching human body temperature, helping to release a smell like rotting meat to attract the flies that pollinate it.

COMMON JASMINE
Jasminum officinale

Jasmine buds and open flowers are used for decoration in Indian weddings.

This species is one of the most strongly scented jasmines, with a rich, sweet honey-like scent. It is likely that it reached Europe from Persia and surrounding areas around the 15th century. Originally, the fresh flowers were collected and embedded in fat to make ointments. It is believed that the scent of jasmine has an aphrodisiac effect, with jasmine flowers used as decor in weddings in parts of India. Scientific studies have shown that the scent used in aromatherapy can lift the mood and increase the feeling of romance.

WELSH POPPY
Papaver cambrica

Welsh poppies will be in full flower by August, seen in both gardens and meadows.

Found in upland parts of Wales as well as other parts of Western Europe, this plant is grown widely as an ornamental in gardens. It thrives in damp, shady places on rocky ground, growing in crevices and can even be found in urban habitats. It is likely to have been part of the Arctic-alpine flora which spread after the glaciers retreated. This unscented bloom will happily self-seed in a garden, filling in gaps and softening paths for a naturalist effect.

SCARLET PIMPERNEL
Anagallis arvensis

This flower only opens when the sun shines, which has led to its
other common name, shepherd's weather glass. The name *Anagallis*
comes from the Greek meaning 'to delight again', which is a reference
to the flower opening each day. In Ancient Greece it was taken as an
antidepressant and in traditional European medicine it was used to
treat various mental conditions. The plant is native to Europe, but has
naturalized around the world, including North America.

CHICORY
Cichorium intybus

According to folklore, chicory can open doors to unseen worlds. There is also a Germanic legend that a blue-eyed woman waiting for the return of her lover was turned into the plant. She waited for him for weeks and then months, while her parents advised her to find someone new. Her response was that she would rather turn into a common wildflower and the belief is that she still waits for him today. The flowers are only obvious in the morning as they close up around midday.

MUSK MALLOW
Malva moschata

Named for its musky smell, this plant grows in fields and on road verges as a wildflower throughout much of Europe and western Asia, but is also a popular garden plant. The Ancient Greeks used it to decorate the graves of friends and it was once used as an aphrodisiac. Often chosen as part of the mix for the creation of a wildflower meadow, it is popular with pollinators. The seed is shaped like a tiny snail shell and covered in minute golden hairs.

BEACH CABBAGE
Scaevola taccada

Found in tropical coastal areas of the Indo-Pacific, this plant is grown along coasts to help prevent erosion of the land from the sea and protect other less tolerant plants from the salt spray of the water. It is also believed that the plant was used for medicinal purposes, including to ease eye irritation suffered by the breath-holding spear fishermen of the Chamorro people in the Mariana Islands.

NIGELLA
Nigella damascena

Above: Nigella is only in flower briefly before each bloom goes over and produces equally attractive seed heads.

Left: The foliage of beach cabbage seen growing here on black volcanic rock in Hawaii.

Also known as love-in-a-mist, these pretty, blue flowers are surrounded by a green ruff, giving the appearance of mist. A popular plant in English cottage gardens since the Elizabethan times, it is native to southern Europe and northern Africa. The seeds are often used in cooking as they give off a peppery scent and they are a favourite with goldfinches. It has also been used in traditional medicine for its antibacterial and antifungal properties.

SEA LAVENDER
Limonium vulgare

Sea lavender is attractive to pollinators, providing a good source of nectar for bees and butterflies such as this tortoiseshell.

Now widely introduced, this plant is native to western Europe and the Azores, found in saltmarshes and coastal flats. While it is not related to lavender, it does have purple flowers and is equally popular with butterflies and other pollinators. The flower spikes are used in dried-flower arrangements as the flowers remain attached even once dry. The flower is believed to represent beauty, sympathy and remembrance.

FEVERFEW
Tanacetum parthenium

The daisy-like flowers bloom throughout the summer months.

The common name of this flower comes from the Latin word *febrifugia*, meaning 'fever reducer'. It is the chemical parthenolide found in the plant's flowers and leaves that has the most health benefits, and can reduce inflammation. It has a long history of use with the Ancient Greeks and early European herbalists. The leaves are ingested fresh or dried, a typical dose being two or three leaves. Believed to be native to the Balkan Peninsula, it was introduced to the USA in the mid-19th century.

Right: Common evening primrose, in a hand-coloured illustration by William Clark from Richard Morris's *Flora Conspicua* (1826).

Opposite top: Arnica, a perennial plant, produces bright flowers typically throughout each July and August.

Opposite bottom: The long, narrow seed heads lend themselves to the common name stork's bill.

COMMON EVENING PRIMROSE
Oenothera biennis

Native to eastern and central North America, this plant is the source of evening-primrose oil. Not related to true primroses (*Primula* spp.), the yellow whorls of flowers do somewhat resemble its namesake. The flowers of many *Oenothera* species open their flowers in the evening, typically to attract moths and evening foraging bees. Most of the plant is edible, and the leaves are sometimes eaten while tender before the flowers have developed.

ARNICA
Arnica montana

This European flower has been used as a medicinal plant for centuries but is now classified as an unsafe herb due to high levels of toxicity, causing stomach problems when taken orally, and skin irritation when applied externally. However, it has been long used as a remedy for bruises and sprains, and many safe products containing arnica extract are available. It is increasingly under threat in its natural habitats as demand for it is greater than supply; in many places it is now a protected plant.

COMMON STORK'S BILL
Erodium cicutarium

This plant has fascinating and specialized seed-dispersal mechanisms, the first of which launches the seeds, with their helix-shaped tails, using stored energy. The second mechanism enables the seeds are able to bury themselves. Through changes in humidity, the tails of the seeds are able to coil and uncoil, creating movement and drilling themselves into the ground.

LACY PHACELIA
Phacelia tanacetifolia

A purple carpet of blooms produced by a mass of lacy phacelia in Oxfordshire, England.

The name *Phacelia* comes from the Greek word meaning 'bundle' and refers to the plant's clusters of scented flowers. Also known as fiddle neck, it is native to the southwest USA and northwest Mexico, most commonly found in deserts. In agriculture it is used as a green manure as it holds onto nitrogen, smothers weeds and is particularly good at attracting bees and hoverflies. It is one of the top honey-producing plants for bees.

PARSLEY
Petroselinum crispum

Parsley, alongside other symbolic foods, is served as part of the festive meal during Passover.

This plant was originally found in and around Greece and the Balkan Peninsula. As with the leaves, which are popular in the culinary world, the flowers are also edible and can be used as a garnish or mixed in with salad. In the Jewish faith, parsley is an emblem of rebirth, symbolizing the initial flourishing of the Israelites in Egypt. During Passover, it is used for Karpas as one of the six symbolic foods served at the Seder, the festive meal.

POTATO VINE
Solanum laxum

The open flowers of the potato vine can be seen fading from purple to white as they mature.

Pretty but poisonous, this plant is related to both the edible potato as well as deadly nightshade, meaning gloves should be worn when handling it. The flowers look very similar to those of the potato plant, but it grows as a climber, and the colour of the flowers changes from pale purple to white over time. Native to South America, it has also naturalized in parts of Australia, and is grown in gardens around the world as an ornamental plant.

HYSSOP
Hyssopus officinalis

The upright flowering spikes of hyssop produce many summer flowers.

A member of the mint family, hyssop is native to southern and eastern Europe. In ancient times it was used to clean temples and other sacred places. It was brought to England in 1597 by John Gerard, the surgeon and apothecary who wrote *The Herball or Generall Historie of Plantes*, one of the world's most famous botanical works. Hyssop soon became featured in many knot gardens. Research in 2002 showed that extracts of the dried leaves were effective in inhibiting the spread of the virus that causes HIV and AIDS; the essential oil also works well as a muscle relaxant.

BEECH DROPS
Epifagus virginiana

Beech drops grow
from the roots
of beech trees on
woodland floors.

This plant grows on the roots of American beech trees and is parasitic, meaning that it doesn't create its own food as most plants do, but instead receives nourishment from its host. Native to North America, beech drops does not cause other plants significant harm. It produces two types of flowers: one that self-fertilizes and another that will cross-pollinate with other plants. The seeds that follow the flowers are small and dispersed by rainwater, germinating only once they pick up on a chemical signal from the root of a beech tree to indicate one is near.

LINSEED
Linum usitatissimum

Above: A farm worker in Northern Ireland collects flax for harvest by hand in the 1950s.

Opposite: Each translucent stem of the ghost plant bears a single flower.

Also known as flaxseed, today this is considered a 'superfood' due to its nutritional benefits, but, it has been grown as a food and for its oil for thousands of years. Linseed contains high levels of omega-3 fatty acids, and is known to support cardiac health, as well as balance hormones during the menopause. Linen is produced from the fibres of this plant – the cloth is noted for being strong and for drying faster than cotton. The flower is used as an emblem for Northern Ireland, the plant being a reminder of the importance of linen as part of its industrial history.

GHOST PLANT
Monotropa uniflora

This plant, which can occur as completely white, with the ghost-like appearance echoed in its common name, is native to temperate areas of North America, northern South America and Asia. It can also be found with black or pink flecks. The reason for its pale appearance is that, unlike most plants, it does not contain chlorophyll. Instead, the plant obtains its nutrients from certain fungi attached to the trees around it. As it does not require sunlight to grow, it can be found in very dark habitats, such as dense forests.

BANANA WATER LILY
Nymphaea mexicana

Native to Mexico and the southern USA, this is known to be among the first water lily species purchased by Claude Monet for his famous ponds. While it is evident through his written orders that Monet was particularly keen on red waterlilies, it was the white and yellow ones that generally proved to be hardier, and tended to do particularly well in his gardens. Monet painted approximately 250 oil paintings of his flower garden at Giverny in northern France, which he created as something 'for the pleasure of the eyes and also for the purpose of having subjects to paint.'

Detail of the Water Lily series by Claude Monet, housed at the Musee de l'Orangerie, gifted to the Prime Minister of France, Georges Clemenceau by Monet in 1918.

MUGWORT
Artemisia vulgaris

A member of the daisy family, this plant is often thought of as a weed. However, the inconspicuous flowers have a history of use that goes back thousands of years. Both the leaves and flowers can be eaten; it is a relative of tarragon with a flavour somewhat like rosemary and sage. Native to Europe, Asia and northern Africa, it is believed that Roman soldiers would put it into their shoes to relieve aching feet. It is also used to try to increase the vividness of dreams, made into a tea or kept near a pillow.

24ᵀᴴ AUGUST

MARSH MALLOW
Althaea officinalis

Native to Asia and Europe, marsh mallow is often found growing in moist soil – as its name suggests. The Ancient Egyptians used the root to make confectionary, and this continued until an alternative method of production using cornflour and gelatine, was developed in the 1800s. The Latin name *Althaea* comes from the Greek word for 'heal', with the leaves and root used for easing pain following childbirth or as an eye ointment.

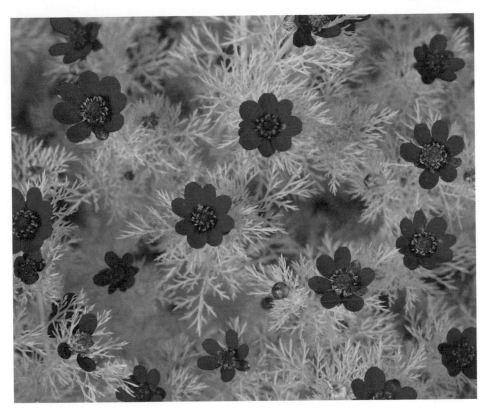

PHEASANT'S EYE
Adonis annua

Above: Pheasant's eye produces vibrant red flowers in the summer.

Opposite top: The flowers of mugwort can be steeped in hot water to make herbal tea.

Opposite bottom: The saucer-shaped flowers of the marsh mallow can usually be seen starting to bloom this month.

Native to Europe, the Mediterranean, northern Africa and western Asia, in the UK pheasant's eye is found in fields and was once considered a weed, but is now thought to be endangered due to intensive farming practices. It can still be found on roadside verges, waste ground and areas where it has been deliberately seeded to try to increase its population. The seeds can remain dormant for a long period in the soil, until a time when germination is more likely to be successful, such as when woodlands are cleared or soil is disturbed.

GREATER BURDOCK
Arctium lappa

This month the purple flowers of burdock can be seen poking out of the prickly ball of bracts they produce.

Originally from Europe and Asia, this plant has become a native weed in places such as North America and Australia. The species name *lappa* comes from the Latin, meaning 'to seize', which refers to the hooked seed heads that catch on walkers and the fur of animals to aid dispersal. In both Western and Chinese traditions it is believed to be a powerful detoxifying medicine. The young stems and roots are edible and are eaten as a vegetable; it is particularly popular in Japan where it is known as *gobo*. Since medieval times burdock has been combined with dandelion to make a drink that is still enjoyed today.

MARGUERITE
Argyranthemum frutescens

Marguerite
flowers appear
like larger
daisies, flowering
continually
throughout
the summer.

Originally from the Canary islands, this plant was used to breed the many marguerite daisies seen in gardens today, either grown as bedding plants or half-hardy shrubs. It is known to have been cultivated in England from the beginning of the 18th century at the Chelsea Physic Garden, and possibly earlier at the Oxford Botanic Garden. The name comes from the old Persian for pearl and it is believed to symbolize purity and innocence.

BABY'S BREATH
Gypsophila paniculata

Baby's breath produces clusters of many small white flowers and is used to celebrate births and weddings.

A member of the carnation family, this plant is native to central and eastern Europe. Tiny white flowers are borne on branching stems, almost like little breaths of cold air, from which its common name is derived. It is popular in floristry as it provides a great background for other flowers and the blooms are long-lasting, which is also why it is a symbol of everlasting love and a reminder of a past love. The flower can be presented to the parents of newborn babies or used at weddings.

CHINESE MEADOW-RUE
Thalictrum delavayi

Chinese meadow-rue is left with many fluffy-looking stamens after the petals have dropped.

Native to China and Myanmar, this is grown as an ornamental plant, and thrives in cooler climates. With its fern-like foliage, it is a popular plant with garden designers, used at the back of borders as it is tall and airy. As the flower's petals drop off, the many stamens are left, looking like balls of fluff or pom-poms. It also has the advantage of not being attractive to slugs, like its relative, the buttercup. After flowering it produces attractive seed heads.

FRENCH LAVENDER
Lavandula stoechas

The purple bracts on top of the inflorescences of French lavender are affectionately known as 'bunny ears' and help distinguish them from English lavender.

Also known as topped lavender, this flower can be distinguished from its English counterpart by the purple tufts on the top of the flower, affectionately known as bunny ears, but more accurately called bracts (modified leaves). While English lavender (*Lavandula augustifolia*) can withstand colder temperatures, French lavender has a stronger scent. The common names refer to the country in which they are most popular, rather than their place of origin. English lavender is native to France, but was used in perfume by English royals; French lavender is native to Spain, but used more commonly in French perfume.

Southern magnolia illustrated in the *Natural History of Carolina, Florida and the Bahama Islands* (1754) by Mark Catesby (who named the illustration *Magnolia altissima*).

SOUTHERN MAGNOLIA
Magnolia grandiflora

With their large white flowers, magnolias take us back to the age of the dinosaurs. The very first flowers looked much like these, and magnolias successfully evolved at a time of ferns and conifers, pollinated by primitive, wingless beetles. As magnolias came about before even bees existed, they didn't and still don't today, produce any nectar. Despite their fragile-looking blooms, this plant has survived and remained virtually unchanged for the past 95 million years.

SNAPDRAGON
Antirrhinum majus

Snapdragons
have a long
flowering season
and will still be
found in bloom
this month.

The common name of this plant comes from the way the flower opens and shuts when squeezed at the back. It is native to southwestern Europe and is a popular garden plant. The Ancient Greeks and Romans believed that the snapdragon provided protection from evil and wore it around the neck. Early German folklore also believed that the flower could protect against evil spirits and people would hang the flowers up near babies.

CRIMSON SCABIOUS
Knautia macedonica

Crimson scabious is nectar-rich, attracting pollinators such as this marbled white butterfly. The cultivar 'Melton Pastels' is shown.

From the honeysuckle family, this plant has pincushion-like flowers, similar to the more common field scabious (*Knautia arvensis*), and is popular with bees and butterflies. Native to southeastern Europe, it is commonly used in wildflower meadow schemes. It produces nectar and pollen-rich flowers as well as seeds that are enjoyed by birds; the shape of the deep burgundy flowers provides a helpful 'landing pad' for visiting insects.

DEADLY NIGHTSHADE
Atropa belladonna

Livia at the deathbed of Augustus, illustration from A Pictorial History of the World's Great Nations by Charlotte M Yonge (c.1880).

Acta est fabula, plaudite
'Have I played the part well? Then applaud as I exit.'

BELIEVED TO BE THE LAST WORDS OF THE ROMAN EMPEROR AUGUSTUS (14 CE)

Deadly nightshade is also known as belladonna, meaning 'beautiful lady', a name that relates to a dangerous practice of women in Renaissance Italy who used the juice of the plant as drops to enlarge the pupils of their eyes so they appeared larger and more striking. A chemical in the plant known as atropine acts as a muscle relaxant, thus allowing the eyes to dilate. It is believed that the Roman empress Livia Drusilla used the juice of the berries either to murder or assist with the suicide of her husband, the Emperor Augustus, who was in ill health.

MEADOW CRANESBILL

Geranium pratense

Meadow cranesbill is one of the showiest wildflowers found in meadows until autumn.

Originally from the Altai Mountains of Central Asia, this is a popular garden plant as it can cope with cold weather. When producing seeds, the stems of this wildflower become erect and generate beak-like pods, which gives the flower its common name. It was once found commonly in hay meadows but due to intense farming practices is now more often found on road verges. In traditional medicine it was used for its antiseptic properties, and as a treatment for cholera and dysentery among other illnesses.

HERB BENNET
Geum urbanum

his plant has become associated with Christianity due to its tri-lobed leaves and five-petalled flowers (believed to represent the five wounds of Christ). In Latin it was known as *herba benedicta*, which later became herb bennet. The plant is more commonly recognized when it has gone to seed, as it creates clusters of dry fruits with small hooks that readily grab onto clothing or the fur of animals as a means of dispersal.

WATER AVENS
Geum rivale

Above: This month is likely the last chance to spot the elegant flowers of water avens.

Left: Both the flower and seed head are charming, but the resilience of herb bennet can make it a nuisance in gardens.

Native to northern and central Europe and North America, this plant is found in damp habitats, such as meadows and woodlands. Traditionally it has been used to treat stomach problems and colds. It is believed that the root can not only be used as a moth repellent, but also boiled up and drunk as a substitute for hot chocolate. Planting water avens is a great way to attract wildlife to the garden, as it provides food for dragonflies, butterflies and bees, and can be used by frogs and newts for shade.

HAREBELL
Campanula rotundifolia

Harebells in
full flower on
Skomer Island,
Scotland, where
they are known
as bluebells.

This resilient wildflower grows on windswept coasts and the bare hill slopes of the temperate parts of the Northern Hemisphere. It is believed to represent childhood and humility, dreaming about harebells is said to symbolize true love. In Scotland it is known as the bluebell, and due to habitat loss is on its way to becoming a threatened plant. The name harebell is said to derive from the flower growing in places frequented by hares. According to folklore, a patch of harebells is a home for fairies, which is why you shouldn't walk through clumps of them.

CURRY PLANT
Helichrysum italicum

Curry plant
growing on the
far right (yellow
flowers) as part
of the vegetable
garden at the
Chateau de
Valmer gardens.

This plant was originally found in and around the Mediterranean. The Ancient Greeks and Romans wove wreaths with the flower heads to decorate their statues, giving the effect of a golden crown. In the 1st century CE, Pliny the Elder used it for its spicy scent and its ability to protect clothes from moths. It is popular in perfumes and a naturally yellow silk is produced by mixing the flower heads with mulberry leaves to feed silkworms; in Sardinia this is used to produce traditional garments. The common name comes from the plant's strong smell.

LADY'S BEDSTRAW
Galium verum

Found in grasslands across the UK, these frothy clusters of bright yellow flowers have a honey-like scent, and were used as bedding before modern mattresses. The plant has a soft and springy quality to it, as well as a bitter scent, which may have deterred fleas. Medieval folklore held that the Virgin Mary gave birth on a bed of lady's bedstraw and bracken. It is said the lady's bedstraw bloomed and the flowers turned from white to gold in honour of the baby Jesus; the bracken, on the other hand, did not recognize the baby Jesus and lost its flowers as a result.

BIRD'S-FOOT TREFOIL
Lotus corniculatus

Above: Bird's-foot trefoil is a common native wildflower that can be found throughout the UK.

Left: Although later in the season, September is one of the best times to spot lady's bedstraw in gardens and grasslands.

This plant is also known as eggs and bacon due to the various yellow and sometimes pinkish tints to the flowers. It is the long seedpods that form after flowering that look like a bird's foot, leading to that part of the common name. The 'trefoil' relates to the shape of the leaves. Native to temperate regions of Eurasia and northern Africa, bird's-foot trefoil is now grown in many other parts of the world and used as food for livestock.

MOROCCAN SEA HOLLY
Eryngium variifolium

The spiky flowers of Moroccan sea holly on top of upright stems can be spotted this month.

Native to northern Africa, this type of sea holly is believed to be a symbol of independence and admiration. While it resembles holly with its spiky leaves, it is not related. The plant provides erosion control in coastal areas as it has a long tap root that holds sand in place. Its striking shape and attractive colour make it a popular garden plant. In the past its roots were candied and eaten as an aphrodisiac.

COMMON TEASEL
Dipsacus fullonum

Looking closely at a common teasel inflorescence reveals the many tiny, purple flowers.

In winter, the spiky heads of teasel provide food for birds like goldfinches, which have to 'tease' out the seeds inside. During the summer months the flower heads are green with rings of purple flowers and enjoyed by the bees. Native to northern Africa, Europe and Asia, this plant was introduced to North America in the 1700s, where it naturalized and became invasive. The word *dipsa* in the Latin name comes from the Greek meaning 'thirst', as the leaves collect water where they meet the stem. A cultivated variety with stronger, curved stems (*Dipsacus sativus*), was used to tease the cloth in early textile manufacture.

YELLOW RATTLE
Rhinanthus minor

A popular annual flower for those looking to establish a wildflower meadow, this plant is semi-parasitic and will help other flowers to grow as it draws water and nutrients from the soil. With fewer nutrients available, the surrounding grasses can no

longer flourish, and more delicate and traditional
species are therefore able to compete and grow through.
Once the flower has gone over, the seedpods ripen and
rattle when shaken, leading to its common name.

Flowering until autumn,
yellow rattle is used to help
new wildflowers establish.

ASTILBE
Astilbe rubra

Astilbe provides late summer colour in gardens, typically flowering until the end of this month.

Also known as false goat's beard, because of the shape of the inflorescences, this fluffy plant is native to eastern Asia, China, Japan and Korea, found growing along the edges of broadleaved forests and shaded rivers. Previously known as *Astilbe chinensis*, it is cultivated around the world as a garden plant, and in the *Journal of the Royal Horticultural Society* in 1902 it was heralded as among the most important hardy perennials to have been introduced. As it requires partial to full shade, it is a useful plant for tricky areas of a shaded garden. It is believed to symbolize patience.

CONEFLOWER
Rudbeckia hirta

Coneflowers produce single-stem blooms in hot colours of yellows, oranges and red.

Native to the central USA, this plant is used in prairie restoration, as it establishes quickly but then fades out to let other longer-lived perennials take hold, which help to stabilize soil and control erosion. It provides good cover and forage for birds as well as butterflies. The flower is popular in gardens as it is later flowering, and also works well with informal gardens styled with grasses. Coneflowers are thought to symbolize encouragement and motivation.

PHLOX
Phlox paniculata

The flowers of phlox are popular in gardens for both their colour and scent.

Popular in gardens around the world, this ornamental plant is native to eastern and central parts of the USA. The name comes from the Greek word for 'flame' and it is said to carry heart-warming sentiments of souls being united. It is believed to have magical qualities and is used in spells for friendship and relationships. It can also represent agreement and understanding.

Vanda sanderiana illustrated in Frederick Sander's Reichenbachia II (1890).

WALING-WALING
Vanda sanderiana

A member of the orchid family, also known as the Queen of Philippines flowers, this plant has spiritual significance to indigenous people of that region. Sometimes classified under the genus *Euanthe*, it is regarded as among the world's most beautiful orchids. Due to over-collecting there have been attempts to make it a national flower in the Philippines to give it greater protection; it is now considered rare in the wild.

BORAGE
Borago officinalis

Borage flowers can be added to salad as attractive edible decorations.

Originally from countries around the Mediterranean, borage can also be found growing in many other European gardens, having been introduced by the Romans. Traditionally it is the leaves of borage that are used as a garnish in the Pimm's Cup cocktail, as they have a refreshing, cucumber-like taste; they can also be used in salads. The star-shaped blue flowers are also edible and can be added to meals and drinks to add flavour and decoration.

LISIANTHUS
Eustoma grandiflorum

Lisianthus bears flowers similar in appearance to its relative the carnation.

Also known as prairie gentian, this member of the carnation family was originally collected from the Texas prairies to be used as an ornamental flower, and is also popular as a cut flower. It was in Japan that plant breeding created the blooms we see today; from the wild plant first known as a Texas bluebell, cultivars of many more colours have been created. It is believed to symbolize appreciation, while the word lisianthus means 'bitter flower', which comes from when the plant was used in traditional medicine and it was noted for having a bitter taste.

ASIAN WATERMEAL
Wolffia globosa

The tiny leaves of Asian watermeal, seen here growing alongside the larger-leaved aquatic plant *Spirodela polyrhiza*.

As the common name suggests, this aquatic plant is native to parts of Asia, but it is also found in parts of the Americas where it is also possibly native. It has been described as the world's smallest flowering plants at 0.1–0.2mm (1/125in) in diameter. It is eaten as part of Thai cuisine and has been recognized for its ability to help clean up waterways as it can absorb excess nutrients. Unusually for a plant, it is also a source of vitamin B12, more often found in bacteria and fungi.

CORNCOCKLE
Agrostemma githago

The delicate flowers of the corncockle each open up to 3–5cm (1½in) wide.

First found growing in cornfields, as the name suggests, corncockle was regarded as a weed found among arable crops, introduced to Britain with agriculture in the Iron Age. Once seed-cleaning techniques improved and pesticide use increased at the beginning of the 20th century, the plant suffered a decline, and is now rare in the wild. Now used in wildflower seed mixes, this plant will flower all summer long and is believed to symbolize duration and gentility.

PASSION FLOWER
Passiflora edulis

The various parts of a passion flower were each given religious meaning.

This exotic-looking plant can be grown in both tropical and non-tropical climates, with some species, notably *Passiflora incarnata*, able to survive in sheltered spots of gardens where temperatures can drop below freezing. It is the species *P. edulis* that produces edible fruit that we know as passion fruit. The plant came to be known as Passion flower after 16th-century Christian missionaries in South America used the flower to explain and symbolize the death of Christ. Among the many plant's characteristics that have been given symbolic meaning, the vine's tendrils are said to resemble whips while the five stamens and three styles represent his wounds and the nails on the cross.

DAHLIA 'BISHOP OF LLANDAFF'
Dahlia 'Bishop of Llandaff'

Dahlia 'Bishop of Llandaff' is one of the most popular dahlia flowers.

Having enjoyed a recent resurgence in popularity, the many shapes and colours of dahlias make them a popular garden plant. Originally from Mexico and Central America, the Aztecs grew the tubers as a food crop. On a trip to Mexico in 1787, French botanist Nicolas-Joseph Thiéry de Menonville wrote about the 'strangely beautiful flowers' he had seen in a garden in Oaxaca; two years later a dahlia was successfully grown in the royal gardens of Madrid, Spain, and the popularity of the plant began to grow.

BELLS OF IRELAND
Moluccella laevis

This plant does not come from Ireland at all, but in fact is native to an area that includes Turkey, Syria and the Caucasus. The reference to Ireland in the common name probably comes from the lush green of the foliage, green being the country's national colour. The flowers are arranged as tall spikes of green bells all the way up the stem, sometimes reaching as tall as 1m (3¼ft). The genus name is derived from the Molucca (Maluku) Islands, in eastern Indonesia where the plant was once thought to be native. The plant is believed to represent good luck.

ROSE CAMPION
Silene coronaria

Above: Rose campion is used by gardeners to provide late summer colour in the garden.

Opposite: Bells of Ireland typically flowers from July to September.

Native to southeastern Europe, rose campion was previously known by the genus name *Lychnis* from the Greek word *lychnos* meaning 'lamp', thought to refer to the felted leaves that were used as lamp wicks. The species name *coronaria* refers to its traditional use in garlands. Its silver-grey foliage and pretty, cerise flowers makes it a popular choice in gardens.

Butterflies such as the monarch butterfly feed upon the nectar from the flowers of the great blue lobelia.

GREAT BLUE LOBELIA
Lobelia siphilitica

This lavender-blue flower is popular with pollinating insects such as butterflies and bees. It has three lower petals which provide a helpful place for bees to land; the bee uses its body weight to open up the flower and crawl inside. If the bee has pollen on its back from a different plant, this will be transferred onto the flower. Some bees simply chew a hole in the base of the flower to take the nectar, avoiding the entrance. The species name *siphilitica* refers to the spurious belief that the plant could cure syphilis.

HATSCHBACH'S FUCHSIA
Fuchsia hatschbachii

Hatschbach's
fuchsia will
flower from late
summer through
to September.

Endemic to parts of Brazil, this fuchsia is popular in temperate regions as it flowers late into the summer and is relatively hardy to cold weather, meaning it can be grown and left outside over winter; many other fuchsias need to be brought indoors or have cuttings taken. As it has a long flowering period, it is a useful and popular plant with pollinating insects as well as gardeners.

VANILLA
Vanilla planifolia

Once the blooms open up, they are pollinated, and then the pods follow on, ready to be picked six to nine months later.

Vanilla is regarded as an essential ingredient for many sweets and treats in Western cooking and baking, but the plant that produces it is far more exotic than its practical use might suggest. Vanilla is a member of the orchid family, native to South and Central America as well as the Caribbean. After its introduction to Europe by the Spanish, it is believed to have been enjoyed by Queen Elizabeth I, which contributed to its extensive popularity. Today, however, it is among the most expensive spices in the world, second only to saffron. Indeed, 99 per cent of vanilla-flavoured products contain synthesized rather than genuine vanilla essence.

EUROPEAN MICHAELMAS DAISY
Aster amellus

European Michaelmas daisies are popular garden plants grown to produce blooms in early autumn once many other plants have finished flowering.

The name *Aster* comes from the Greek for 'star', referring to the shape of the flowers. This plant is popular in gardens for its late summer and autumn blooms, and is also enjoyed by many pollinating insects. The flowers are in bloom around Michaelmas Day, a Christian celebration originally falling on 11 October, but now celebrated on 29 September. It is the feast day of Saint Michael the Archangel and All Angels, associated with protection as the days get shorter and colder.

WILD TOBACCO
Nicotiana rustica

French postage stamp (1961) depicting French diplomat Jean Nicot as well as the flowers and leaves of the nicotine plant.

While other species of *Nicotiana* are found more commonly in gardens, this plant is more potent, containing up to nine times more nicotine than many other varieties. Known as *mapacho* in South America, it is used as a psychoactive substance by shamans. In some parts of Vietnam it is smoked after a meal to aid digestion. Its genus name comes from the French diplomat and scholar, Jean Nicot de Villemain, who introduced tobacco to the French court in the 16th century.

KING PROTEA
Protea cynaroides

Prized for its impressive display in floristry, the 'flower head' of the king protea is actually made up of many small flowers.

The national flower of South Africa, this species has the largest flower head of all the proteas. The flowers in this genus come in a variety of colours and sizes, which is why they were named after the Greek god Proteus, who could change his shape at will. Proteas are popular in floral displays and arrangements, as not only are they impressive and exotic-looking, but they remain open for a long time, with a good vase life.

In Hans Christian Andersen's fairy tale, Thumbelina is saved by a swallow and meets the fairy flower prince. Around him is a plant with large white flowers and broad leaves, matching the description of Angel's Trumpet, here illustrated in watercolour by Eleanor Vere Boyle (1872).

ANGEL'S TRUMPET
Brugmansia arborea

Strongly fragrant, trumpet-shaped flowers adorn this small tree, making it popular in gardens. Native to South America, where it is now extinct in the wild, it can still be found in gardens or naturalized, especially in Asia. It is a striking plant, and while it can tolerate temperatures as low as 10°C (50°F) in sheltered gardens, elsewhere it needs to be brought inside over winter.

NETTLE
Urtica dioica

Stinging nettles will still be in flower this month – it is recommended to not harvest, cook and eat them when in flower as they have a laxative effect.

The yellow-green flowers of the common stinging nettle hang like catkins, dropping down from the stems. The plant itself is hugely beneficial to wildlife, providing a food source for small tortoiseshell and peacock butterflies. When the flowers are over, the seeds are eaten by birds and can be collected and eaten by humans. They contain omega-3 fatty acids, and historically were fed to horses, to improve both their glossy manes and energy levels.

FORGET-ME-NOT
Myosotis scorpioides

Forget-me-not are found throughout the UK, typically flowering from May through to October.

The name *Myosotis* means 'mouse ear' in Greek and describes the shape of the small leaves of this plant. It is believed to be a symbol of remembrance and fond memories; in the Netherlands people sometimes give mourners attending funerals seeds of forget-me-nots to plant at home in honour of the deceased. The flower is also meant to symbolize true love and loyalty, as well as secrets to be remembered.

CARNATION
Dianthus caryophyllus

Oscar Wilde wearing a green carnation, photographed in 1899 by W. & D. Downey.

The botanical name of this flower comes from the Greek meaning 'flower of the gods', and while many people call carnations 'pinks', this does not relate to the colour of the petals, but rather the ruffled shape that resembles the jagged edges in fabric made when cut with pinking shears. Carnations have been grown extensively over the last 2,000 years, which makes it difficult to know where they originated, but it is thought they first came from the Mediterranean. These flowers are worn as buttonholes, and in 1892 the green carnation became a queer symbol when Oscar Wilde instructed his friends to wear them to the opening night of his comedy *Lady Windemere's Fan*.

CATNIP
Nepeta cataria

Catnip is a culinary herb that is also attractive to cats. It has a long flowering season.

Native to Europe, the Middle East as well as southwestern and Central Asia, this plant has also become naturalized and grows wild in other parts of the world, including North America. Alongside some other *Nepeta* species, catnip affects over two-thirds of domestic cats, making them feel blissful or aggressive. The plant releases a volatile chemical that stimulates the brain, creating a sexual response, which is why kittens are not affected.

COMMON ZINNIA
Zinnia elegans

Zinna flowers provide colour later in summer, even into a mild October.

Native to Mexico, but widespread in gardens where they are easy to grow and bloom over a long period, this flower is adored by butterflies and other pollinators. Traditionally they have been given as a going-away present or gifted to a friend who hasn't been seen for some time, representing affection. When first introduced to Europe, it is believed that zinnias were called 'poorhouse flowers' as they were so common and easy to grow.

STEPHANOTIS
Marsdenia floribunda

Stephanotis is popular as a house plant, and when cared for well it will still produce blooms this month.

Popular as a houseplant for its fragrant blooms, this plant is native to the island of Madagascar off the coast of East Africa. The common name derives from the Greek *stephanos* for 'crown' and *otis* for 'ear', which refers to the arrangement of the stamens and the ear-shaped petals. Other common names include Madagascar jasmine and bridal wreath; indeed, stephanotis is a popular choice in floral bouquets, as it is thought to be a symbol of marital happiness. The flowers need to be deadheaded as soon as they go over as the heavily scented blooms turn sour.

SMALL SCABIOUS
Scabiosa columbaria

Small scabious flowers for a long period and provides nectar and pollen for a wide range of insects.

Found in parts of Africa and Europe, its common and Latin names derive from its history as a treatment for scabies by the Romans. This meadow wildflower is popular in gardens, and the alternative common name of dwarf pincushion flower comes from its cushion-like centre, and stamens that stick ou,t resembling pins. The blooms are particularly attractive to butterflies and are commonly grown as cut flowers due to their long stems.

CAPE PRIMROSE

Streptocarpus saxorum

The final flowers of Cape primrose will be produced in the Northern hemisphere this month, and just be about to start flowering in South Africa.

Native to parts of Africa, streptocarpus is a popular houseplant. The name derives from the Greek *streptos* meaning 'twisted', and *karpos* meaning 'fruit', as the seedpods are long and twisted. In 1818, English plant collector James Bowie sent seeds of *Streptocarpus* over to the Royal Botanical Gardens, Kew; this marked the beginning of much cultivation and breeding, creating the many incredible colours and varieties we see today.

POT MARIGOLD
Calendula officinalis

Pot marigold
grown as a
companion
plant in this
vegetable garden
to deter pests.

Thought to be native to southern Europe and the eastern Mediterranean, this plant has been cultivated for so long that no one is sure of its origin. The flowers themselves are edible and can be added to dishes as a garnish, and have also been used to make dye for fabrics and tea. It is used by gardeners in vegetable gardens as it is thought to deter pests, while attracting pollinators at the same time.

295

AUTUMN CROCUS
Colchicum autumnale

The pretty purple flowers of autumn crocus begin to emerge this month.

Resembling the spring crocus, but flowering in autumn, these flowers emerge and die back before the leaves appear, leading to the alternative common name, naked ladies. Found in woodlands, hay meadows and gardens in Europe and New Zealand, and also known as meadow saffron, the plant must not be mistaken for the true crocus from which saffron is derived, as all parts are dangerous if ingested by humans and animals. However, the chemical that makes the plant poisonous, colchicine, has been used to treat gout.

SUN SPURGE
Euphorbia helioscopia

Sun splurge will typically flower from May to October in the UK.

Native to many parts of the globe, including much of Europe, Asia and northern Africa, sun spurge has also been called mad woman's milk due to the plant's toxic, milky sap. This sap was traditionally used to cure warts but is an irritant, causing photosensitive skin reactions and is poisonous if ingested. In some parts of its range, such as the lowlands of Scotland, it is becoming less common as it prefers arable fields which are in decline.

PURPLE CONEFLOWER
Echinacea purpurea

The common name coneflower comes from the shape of the central disc of each bloom.

Native to North America, this is a relative of the sunflower. The word *echinacea* is derived from the Greek for 'spiny one', as the plant is said to resemble a sea urchin. Traditionally, Native American people used the plant to treat wounds such as burns and insect bites, as well as coughs and stomach cramps. Research has shown that the plant can stimulate the immune system, as well as containing anti-inflammatory properties.

FIELD SCABIOUS
Knautia arvensis

Field scabious typically flowers from July to October.

Each of these plants can produce up to 50 flowers, making it a bounty for wildlife, with seed-eating birds like finches and linnets in particular enjoying the seeds. The name probably comes from the rough and hairy texture of the stem resembling scabby skin. Herbalists from the time of antiquity would treat ailments using plants that resembled the body part associated with the illness, a belief referred to as the doctrine of signatures; thus field scabious was used to treat scabies and manage itching.

Kiss-me-over-the-garden-gate depicted in a hand-coloured etching from Pierre Joseph Buchoz' *Collection precieuse et enluminee des fleurs les plus belles et les plus curieuses, qui se cultivent tant dans les jardins de la Chine, que dans ceux de l'Europe* (1776).

KISS-ME-OVER-THE-GARDEN-GATE
Persicaria orientalis

Thought to originate from China or Uzbekistan, this plant has been particularly popular in American gardens, with President Thomas Jefferson supposedly a big fan of the bold flowers. It is also known as princess-feather, the name perhaps referring to the long, soft arching spikes of pink blooms. It attracts hummingbirds among other pollinators, and self-seeds readily, which decreases its attractiveness for some.

APPLE OF PERU
Nicandra physalodes

Both the dark buds and open flowers appear simultaneously on the apple of Peru plant, which is named for its decorative fruit.

From the Nightshade family, this plant is thought to have originated from western South America, from countries such as Peru. It is grown in gardens for its attractive flowers and unusual fruits, which are sometimes used dried in floral arrangements. It can be found in gardens, having been seeded unintentionally from bird-seed feeders, but the plant does not survive cold weather. In some parts of the world, it is an agricultural weed, but it is also grown in glasshouses to discourage whiteflies – leading to another common name, the shoo-fly plant.

WILD PETUNIA
Ruellia humilis

This low-growing plant produces an abundance of flowers.

Native to eastern parts of the USA, the wild petunia is found growing in prairies, fields and dryish open woodlands. The species name *humilis* refers to the low-growing nature of the plant, and it is increasingly being used in gardens in North America as a native plant to encourage wildlife. In the UK, the plant's ability to cope with dry conditions makes it popular with gardeners. The genus name honours Jean Ruelle (1474–1537), a French herbalist and physician to Francis I, King of France.

BUTTERFLY BUSH
Rotheca myricoides

Not to be confused with buddleja (see page 315), this butterfly bush starts flowering this month in Botswana.

While this common name is also applied to *Buddleja*, which attracts butterflies, this flower is truly deserving of the name, as the violet blooms form butterfly-like shapes. Each flower comprises four light blue side petals that look like wings; a fifth dark blue lower petal which resembles a butterfly's head, thorax and abdomen; finally there are flamboyant, curved stamens like antennae. The flowers are followed by black, fleshy fruit. Originally from Africa, the butterfly bush is now grown around the world.

IVY-LEAVED CYCLAMEN
Cyclamen hederifolium

Ivy-leaved cyclamen are autumn flowering and can be spotted this month in gardens and woodlands.

Native to woodland and rocky areas of the central and eastern Mediterranean, both the common and Latin names of this plant refer to the similarity of the foliage to that of ivy. Growing from tubers, the flower stalks will often twist after flowering, bringing the seed heads closer to the ground, helping with dispersal and germination. It is believed that the Ancient Greeks used the tubers to make cakes that were used as a love potion. It was also believed to cure baldness and speed up labour.

COMMON MORNING GLORY
Ipomoea purpurea

Common morning glory flowers into October, making it a popular plant with gardeners.

Native to Mexico and Central America, common morning glory is regarded as a weed in many parts of the world where it was introduced as a garden plant. It bears a profusion of flowers which quickly go over, but while in bloom new flowers open every morning as the sun rises, leading to its common name. The seeds are believed to contain d-lysergic acid amide (LSA), which if consumed can cause psychedelic hallucinations similar to the drug LSD.

LEMON BALM
Melissa officinalis

Lemon balm produces small, white flowers from June through to October.

A member of the mint family, this plant is now widely naturalized but was originally native to parts of Europe, Asia, Iran and the Mediterranean. It is very popular with bees and is used in the production of honey. It has been used as a medicine since the 10th century, and more recently clinical trials have proved it is effective in aiding sleep and relieving stress. The leaves can be steeped in hot water to be drunk as a tea, but flowering stems can also be added to baths for a floral soak.

The lobster claw flowers are popular with birds such as this coppery-headed emerald hummingbird.

LOBSTER CLAW
Heliconia rostrata

Native to southern Central America and northern South America, unlike other *Heliconia* species, this plant has downward-facing flowers. Upright flowers are able to store water for birds and insects, but when they face downwards they can offer nectar to birds. This is a popular plant in tropical gardens and attracts hummingbirds. The name lobster claw refers to the beak-like bracts around the flowers, the bright colours attracting pollinators.

WHITE GAURA
Gaura lindheimeri

Native to Texas and southern Louisiana in the USA, this is a popular garden plant in temperate regions around the world. The cultivar 'Whirling Butterflies' is especially popular, with many delicate flowers that dance around in the wind on tall stems, much like the butterflies that visit the plant. In the wild, it is found in prairies and pine forests, while its resilience to some degree of drought makes it popular with gardeners.

SELFHEAL
Prunella vulgaris

Above: Selfheal can be steeped in hot water to make a herbal tea.

Left: White Gaura 'Whirling Butterflies' is a popular cultivar of this heavily flowering plant.

Native to much of the temperate world, this member of the mint family has a long flowering period, producing blooms ranging from blue to white or pink. In many places it has become invasive and is treated as a weed, however, the whole plant is edible and has a long history of medical use. It was used to alleviate skin irritation, including nettle stings, and made into a tea to treat problems such as fever and heart problems among others. Research has shown that the plant does indeed have antioxidant and anti-inflammatory properties as it contains ursolic acid.

CHINESE HIBISCUS
Hibiscus rosa-sinensis

A statue of the
Hindu goddess
Kali in the
Sri Veerama-
kaliamman temple
in Singapore.

Grown as an ornamental flower, Chinese hibiscus is eaten in salads on the islands of the Pacific and has many other uses. In India, it is used to shine shoes, and the red variety is used to worship Kali, the goddess of time, doomsday and death. In China, it is believed to have medicinal properties, and was traditionally used to treat dysentery and diarrhoea. Since 1960 it has been the national flower of Malaysia, where its name means 'celebratory flower', and it features on all bank notes.

STRAWBERRY TREE
Arbutus unedo

The strawberry tree produces both flowers and the resulting fruits simultaneously, the latter thought to resemble strawberries.

Known for the edible berries that bear some resemblance to strawberries, each fruit from the strawberry tree takes 12 months to mature; the redder and riper it is, the sweeter it tastes. The species name *unedo* is thought to come from Pliny the Elder, who is believed to have said of the fruit, *unum tantum edo*, which translates as, 'I eat only one'. It is unknown whether this is because the fruit tasted so good that one was sufficient, or that he did not care for it and therefore did not want to eat any more.

BLACKTHORN
Prunus spinosa

While the blackthorn flowers in spring, the sloes develop later and can be collected this month to flavour gin.

Often confused with the similar-looking hawthorn, one way to identify blackthorn is to spot the presence of flowers before the leaves start to show – the hawthorn blooms after its leaves have emerged. It is from this plant that the dark fruit known as sloes are collected to make sloe gin and preserves. As an early-flowering shrub, with both male and female parts produced on the same flower, it provides a valuable source of pollen and nectar for bees.

VERBENA
Verbena bonariensis

Verbena adds height to the border and self-seeds freely.

Also known as purpletop, this plant is actually native to tropical areas of South America. The name verbena is derived from the Latin for 'sacred bough', which references the leafy twigs of vervain (*Verbena officinalis*) that were carried by priests and used in medicine. It is a popular drought-tolerant, wildlife-friendly garden plant. In Christianity, it is believed that verbena was used to treat Jesus's wounds after he was taken down from the cross, symbolizing protection, healing and happiness.

ST JOHN'S WORT
Hypericum perforatum

From left to right: Broadleaf plantain, stinging nettles and St John's wort here being dried to be used for medicinary purposes.

This plant is used in traditional herbal medicine to help treat depression, and in some instances has been found to be as effective as standard antidepressant medications. It may also be useful in alleviating some symptoms of menopause, but it is important to note that it can compromise the effectiveness of many medications, including heart medication and birth-control pills.

BUTTERFLY BUSH
Buddleja davidii

True to its
name, the
butterfly
bush attracts
pollinators such
as this peacock
butterfly.

Eleven species of caterpillar from the butterfly and moth families are known to feed on the leaves and flowers of this plant, which explains the many butterflies that can be found swarming around its flowers during the summer. Native to regions of Japan and China, it was introduced to Britain in the late 1800s, and began to colonize wasteland in the 1930s. Today it can be spotted along railway tracks and other disturbed sites, and is very useful for wildlife as a source of nectar.

BIRD OF PARADISE
Strelitzia reginae

In South Africa, November sees the bird of paradise in bloom.

This spectacular flower is indigenous to South Africa but is grown in warm climates around the world for its exotic and ornamental appeal. The flower resembles a brightly coloured bird, but its mechanics that allow for pollination are equally impressive. Sunbirds perch on the blue petals, which are fused together with a nectary at the base. As they reach down to feed, hidden anthers within the petals brush against the birds' feet, covering them with pollen. This process is repeated when the bird visits the next flower and pollination takes place.

COMMON TOADFLAX
Linaria vulgaris

Common
toadflax is a
food source
for pollinating
insects, including
this marbled
white butterfly

Originally from temperate regions of Europe and Asia, this plant is mainly found in open grassland or on dry soils. The yellow flowers are similar to snapdragons and require insects to be heavy enough to open the flowers and pollinate them; consequently they are a food source for bumblebees and honey bees as well as moths and butterflies. The name supposedly refers to the way the head of the flower resembles a toad's mouth, but also because it was thought that toads would shelter among the stems of the plant.

SAFFRON CROCUS
Crocus sativus

The long, red pollen-bearing stigmas that are harvested as saffron can be seen protruding from the centre of each flower.

In the centre of these pale purple flowers are the pollen-bearing stigmas that are harvested for use as the spice and dye known as saffron. While cultivated mostly in Iran, saffron is also grown for commercial use in France, Spain and parts of Italy too. The three stigmas are hand-picked from each flower and dried to create the pungent, sweet-smelling spice. Dishes such as bouillabaisse and paella rely on saffron, but with around 50 flowers required to produce just one tablespoon of saffron, it is the most expensive spice in the world.

COMMON IVY
Hedera helix

Dionysus shown with fruit and ivy in his hair from the Mosaic of Dionysus, in the ruins of Corinth, Greece, second half of the 2nd century BCE.

Native to most of Europe, common ivy has been introduced to many parts of the world as an ornamental plant, and is very wildlife-friendly, providing both a food source and habitat. Ivy was associated with Dionysus, the Ancient Greek god of wine, pleasure and excess, who was often depicted wearing a crown of ivy, as it was believed to grow over the mythical mountain of Nysa, his childhood home. In the Middle Ages, ivy berries hung outside a tavern indicated that wine was sold inside.

HEATHER
Calluna vulgaris

A Yorkshire moor in England lit up with a carpet of heather in full bloom. The flowers will continue to open into early November.

Native to Europe and parts of temperate Asia, this plant is found mostly in heathlands and moorlands. White heather, which is less common than the purple variety, is considered lucky in Scotland. It is believed to be found only on ground where blood has not been shed in battle, or on the graves of fairies. Queen Victoria brought sprigs of white heather back from Balmoral to England, popularizing the superstition. Sprigs are sometimes put inside bridal bouquets for good fortune.

Common dandelion depicted in a chromo-lithograph after a botanical illustration by Walther Müller from *Hermann Adolph Koehler's Medicinal Plants* (1887).

DANDELION
Taraxacum officinale

Found nearly everywhere around the world, after they have flowered dandelions create seed heads that are extremely successful at dispersal, as they are able to cover large distances in the wind. The seeds, with their parachute-like appendages, respond to moisture in the air. This allows them to hold onto the plant until the weather is dry enough for successful dispersal. The flowers used to be known as 'fairy clocks', as it was thought that the number of breaths it took to disperse the seeds told the time, or that a wish would be granted.

FIRECRACKER BUSH
Bouvardia ternifolia

The firecracker bush is also known as the hummingbird flower due to birds, alongside bees and butterflies that enjoy its nectar-rich flowers.

Grown for its bright, red flowers, this plant can be found in parts of the southwestern states of the USA, much of Mexico and south into Honduras. The flowers are pollinated by hummingbirds that feed on the nectar. The Spanish name for the plant is *trompetilla* meaning 'little trumpet', due to the shape of the flowers. This charming bloom is also believed to symbolize enthusiasm and a zest for life.

KANGAROO PAW
Anigozanthos rufus

Above: The furry red flowers of the kangaroo paw are thought to resemble its namesake.

Opposite top: This flower is named for the city near Perth in Australia from where it originates.

Opposite bottom: The name refers to the dark centre to each flower.

Native to the southern coasts of Western Australia, the flowers of this plant are tubular with a velvety texture, resembling a kangaroo's paw. It is also the emblem of Western Australia. Used in traditional medicine to heal wounds, it is now included in some skincare products as it is believed to have regenerating properties that help reduce lines and wrinkles. Due to the colour and architectural property of the plant, it is popular in gardens of warmer climates and used as a cut flower.

GERALDTON WAX FLOWER
Chamelaucium uncinatum

Endemic to southwestern Australia, this plant is grown commercially for the cut-flower industry as the flowers can last up to three weeks. They are known as wax flowers due to the slightly waxy feel of the petals, which helps them last longer than most other blooms. With a slightly lemony scent, they are said to symbolize patience and lasting love, making them popular for use in floral crowns and bouquets at weddings.

BLACK-EYED SUSAN VINE
Thunbergia alata

Native to East Africa, the black-eyed Susan vine is a popular garden plant grown as a tender annual in many cooler parts of the world. In warmer areas, such as parts of North America and Australia, this fast-growing vine has escaped from gardens and become invasive. In East Africa, it is used for animal feeds and to treat skin problems, joint pains and eye inflammation.

Common Poppy
Papaver rhoeas

In Flanders fields the poppies blow
Between the crosses, row on row,
That mark our place; and in the sky
The larks, still bravely singing, fly
Scarce heard amid the guns below.

We are the Dead. Short days ago
We lived, felt dawn, saw sunset glow,
Loved and were loved, and now we lie,
In Flanders fields.

From 'In Flanders Fields' by
John McCrae (1915)

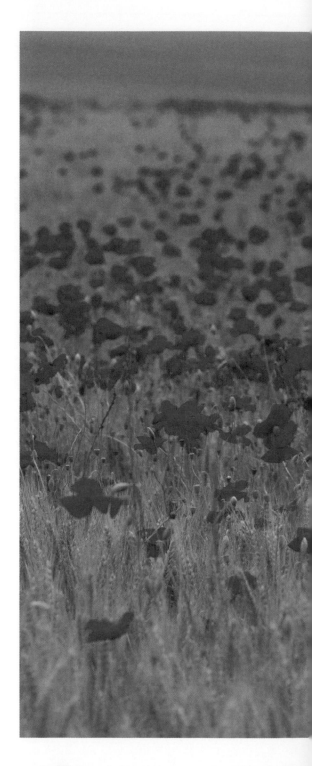

Poppies are used to remember those who have given their lives in battle or died on behalf of their country. After the First World War, the muddy battlefields were sites of death and destruction, and it looked unlikely that anything would grow there. However, wild poppies thrived in the disturbed conditions of the soil and came to represent resilience and hope, with the red petals a reminder of bloodshed. The Canadian doctor Lieutenant Colonel John McCrae's poem records this and inspired many others to use the poppy as a symbol of remembrance.

A sea of poppies in the fields of a World War I battlefield in the Somme, France. While the flowers will have now gone over, many poppy symbols will be seen worn on this day in remembrance.

BLUE ROCK BINDWEED
Convolvulus sabatius

Blue rock bindweed produces clusters of funnel-shaped, purple flowers.

Also known as *Convolvulus mauritanicus*, this plant is native to coastal regions of northwestern Africa, such as Algeria and Morocco, as well as southern Italy; the species name *sabatius* refers to the Savona region of Italy where it can be found. The plant is often seen growing in gardens outside its native range, and is treated as an annual plant grown for the summer. Despite being a relative of the infamous and invasive bindweed, this is a much less vigorous plant in most conditions.

WHITE DEAD-NETTLE
Lamium album

White dead-nettle produces blooms over a long flowering period, even into November.

Native to Europe and Asia, the common name refers to the similarity of the plant to the stinging nettle (*Urtica dioica*), but without the sting. Once in flower, white dead-nettle looks markedly different with its distinctive white blooms. It is popular with bees and is very wildlife-friendly as it blooms for much of the year: the flowers provide food for various bees and moths; while the leaves are eaten by caterpillars of the garden tiger moth and green tortoise beetles.

GOLDEN WATTLE
Acacia pycnantha

Golden wattle represented on the coronation dress of Queen Elizabeth II, photographed here with her maids of honour, by Cecil Beaton (1953).

Native to southeastern Australia, this plant is grown in gardens for its bright yellow, fragrant flowers. Golden wattle is often the first plant to germinate after a bushfire and is therefore seen as a symbol of resilience and rebirth. It is the floral emblem of Australia and was used to represent Australia on the white satin coronation gown of Queen Elizabeth II in 1953, embroidered with yellow wool for the flowers and green and gold thread for the foliage.

Gentian sage shown (second from left) in a chromo-lithograph after an illustration by the author from Mrs Jane Loudon's *Ladies Flower Garden of Ornamental Perennials* (1849).

PL 83.

1 Salvia fulgens __2 Salvia patens __ 3 Salvia Grahami __ 4 Salvia angustifolia

GENTIAN SAGE
Salvia patens

Native to parts of central Mexico, this plant is adored by gardeners for its blue flowers, a less common colour in the horticultural world. It was introduced to gardens the UK in 1838, and was popularized by Irish gardens and botanist William Robinson, who praised the plant in the 1933 edition of his famous book *The English Flower Garden* as 'doubtless, the most brilliant in colour, being surpassed by none and equalled by few flowers in cultivation'.

TATARIAN ASTER
Aster tataricus

This plant, native to Siberia, northern China, Mongolia, Korea and Japan, has been one of the fundamental herbs of traditional Chinese medicine for at least 2,000 years due to its antibacterial properties. More recently, however, it has been discovered that these properties do not occur in the plant itself. The active ingredients, known as astins, which are now being explored by cancer researchers are actually the product of a fungus, *Cyanodermella asteris*, that exists inside the plant.

DARK SKY-BLUE ALOHA
Nemesia caerulea

Native to southwestern South Africa, where it grows on exposed slopes, this plant produces small, delicately scented flowers in shades of blue, pink and white. It is a compact plant, adapted to survive windy conditions, making it popular with gardeners for use in bedding and containers. Believed to be a symbol of friendship, the many cultivars bred from the natural species of *Nemesia* are available in a wide range of colours.

Right: Ragwort illustrated in *The Aurelian. A Natural History of English Moths and Butterflies* by Moses Harris (1840).

Opposite top: The Tatarian aster is vigorous and will flower from late summer to frost, typically into November; the cultivar Aster 'Jindai' is shown here.

From the purplish-blue flowers of the dark sky-blue aloha flower, many nemesias in other colours have since been bred and proven popular with gardeners.

RAGWORT
Jacobaea vulgaris

This wildflower is known to provide problems for grazing animals such as horses as it contains toxic alkaloids that, when eaten, can lead to liver failure and death. Originally native to Eurasia it is now found more widely. As long as gardens are more than 50m (164ft) from grazing land, this wildflower can harmlessly provide a food source for wildlife: it is an important nectar source for many insects, attracting many butterflies and moths. It is also known as mare's fart due to the vile smell given off by the crushed leaves.

ASHWAGANDHA
Withania somnifera

The attractive orange fruits of the ashwagandha each enclosed in a calyx, follow after the small green flowers have gone over.

Also known as Indian ginseng, this member of the nightshade family is found growing in India, parts of Africa and the Middle East. The bell-shaped flowers are small and green, and are followed by dark orange fruit. The species name *somnifera* comes from the Latin meaning 'sleep inducing' and is used in Ayurvedic medicine as well as various cultures as a traditional medicine, believed to lower stress levels and anxiety.

AUTUMN SNOWFLAKE
Acis autumnalis

Starting to flower in early autumn, autumn snowflake will typically flower into November.

This dainty bulb is native to areas around the western Mediterranean including Algeria, Morocco, Portugal and Spain. The bell-shaped flowers are produced after the leaves have died down, which makes the flowers easier to spot. Despite its delicate appearance, the autumn snowflake is a hardy plant that withstands the cool temperature of the later part of the year. New leaves emerge again in spring once the flower has gone over.

BLUE TWEEDIA
Tweedia coerulea

This tropical twining vine is enjoyed for the vibrant blue flowers that it produces.

Also known as *Oxypetalum coeruleum*, this plant is originally from southern Brazil to Uruguay, where it is mostly found in the rocky areas. The genus name *Tweedia* honours a 19th-century head gardener, James Tweedie of the Royal Botanic Gardens, Edinburgh, who undertook expeditions to South America and brought many plants to the attention of the wider world.

GERBERA DAISY
Gerbera jamesonii

Gerberas are now produced in an assortment of colours, including these from the 'Mega Revolution' series.

The Ancient Egyptians believed gerberas symbolized a closeness to nature and devotion to the sun as they track its movement during the day. Native to southeastern Africa, the cheery colours of these showy blooms, with a single flower on each strong stem, makes them highly suitable for use in floristry and they are the fifth most popular cut flower in the world, after roses, carnations, chrysanthemums and tulips.

MEXICAN SUNFLOWER
Tithonia diversifolia

Also known as tree marigold, the bushy plant produces large daisy-like flower heads.

Native to Mexico and Central America, this plant has been introduced around the world for ornamental purposes. However, it can be used to provide more than just colour in the garden as it has been trialled as a cheaper and more environmentally friendly alternative to chemical fertilizers. During trials in western Kenya, it was found that *Tithonia diversifolia* could be planted as a companion crop for maize, providing green manure to improve soil fertility and thus increase productivity.

CHRYSANTHEMUM
Chrysanthemum spp.

Women prepare chrysanthemums for show in Tokyo, Japan. Photograph taken by Herbert Ponting (c.1907).

Native to East Asia and northeastern Europe, the name of this flower literally translates from Ancient Greek as 'gold flower'. In China, the tradition of breeding and growing chrysanthemums dates back 1,600 years, and festivals to celebrate it are held every year, as well as in Japan. It was believed that the plant could bring long and healthy life, so teas and wines are traditionally made from the flowers. It is also believed to represent happiness, love and longevity.

COMMON VETCH
Vicia sativa

Common vetch can flower even as late as November when weather is mild, providing nectar for pollinators such as the common blue butterfly.

With sweet-pea-like flowers, this plant is grown as fodder for livestock, but archaeological evidence shows they were also eaten by humans: the remains of the plant have been found in early Neolithic sites in the eastern Mediterranean. Not only is common vetch attractive to pollinators, but it has developed a mechanism to defend itself: the plant produces nectar in stipules along its stem, as well as in the flowers, encouraging the presence of ants which then discourage pests and larger predators.

OLEASTER
Elaeagnus × submacrophylla

Starting to form in autumn, the small fragrant flowers of the oleaster open and can be spotted along stems this month.

This plant is grown for its attractive foliage, making for a versatile shrub or hedge. However, from autumn into late winter it also produces white flowers that are sweetly fragrant. The plant itself is a cross between the two plants *Elaeagnus macrophylla* and *Elaeagnus pungens*, which both originate from Asia. Also known as silverberry, silvery orange fruits follow the flowers, which are edible but astringent until fully ripe.

BIG BLUE LILYTURF
Liriope muscari

Big blue lilyturf flowers from late summer into November with attractive upright purple flowers.

With flowers that resemble grape hyacinths, this popular garden plant produces spikes of lavender flowers from late summer until autumn. Native to China and Japan, the flowers are followed by dark berries. The name lilyturf references the evergreen, grass-like foliage, which also contributes to its value in the garden. The genus name references the Greek woodland nymph Liriope, who was the mother of Narcissus in Ovid's *Metamorphoses*.

ANCHOR PLANT
Colletia paradoxa

The anchor plant produces tiny white, scented flowers this month along the modified stems.

Originally from temperate South America, this plant is grown in winter gardens in the Northern Hemisphere as it can grow well outdoors in sheltered areas. It produces white, bell-shaped flowers, reminiscent of heather, but also flat shoots in place of leaves. These shoots are triangle-shaped, almost thorn-like, lending themselves to the common name anchor plant. The flowers are scented and produced in tight clusters.

GLOSSY ABELIA
Linnaea × grandiflora

Glossy abelia produces pale pink, fragrant blooms which will flower from mid-summer to as late as November in mild conditions.

Also known as *Abelia × grandiflora*, this late-flowering plant is the hybrid of two honeysuckle species, *Lonicera chinensis* and *L. uniflora*. It was created in 1866 at the Rovelli nursery in Italy. The species name *grandiflora* comes from the Latin meaning 'abundant flowers', which this multi-stemmed shrub displays late in the year. The fragrant blooms are borne in clusters on arching stems.

JAPANESE ONION
Allium thunbergii 'Ozawa'

Still flowering once many other plants have gone over, the Japanese onion produces clusters of purple blooms.

The species plant is native to Japan, Korea and coastal China where it can be found growing in woodland margins. It is a useful garden addition as it typically flowers after most other plants, usually between September and November. The flowers are produced in globular clusters, with grass-like leaves that have the characteristic onion smell of an allium when crushed. If seen in a garden, it is quite likely to be the cultivar 'Ozawa', as this is widely available commercially.

WINTER-FLOWERING CHERRY
Prunus subhirtella 'Autumnalis Rosea'

From around the beginning of this moth, winter-flowering cherry produces semi-double blooms, which means they have two to three times more petals than single flowers, but not as many as double blooms.

The species plant of *Prunus subhirtella* is native to Japan, with the cultivar 'Autumnalis Rosea' producing semi-double blooms from November, followed by even more blooms in spring. After the flowers, it produces attractive autumn foliage as well as berries that are popular with birds. This plant flowers from a young age and is smaller than the wild species.

GARDEN LOBELIA
Lobelia erinus

Garden lobelia, shown here in a hanging basket, will produce blooms until the first severe frosts.

With flowers produced in blue, pink, violet or white, this member of the bellflower family is native to southern Africa. Found on lower mountain slopes and coastal flats, the plant produces a profusion of flowers. It is popular as a half-hardy annual, grown in temperate countries where it is used to trail out of hanging baskets or to edge borders.

LAURUSTINUS
Viburnum tinus

Above: *Viburnum tinus* flowers this month and until spring, with blooms followed by attractive, metallic blue fruits.

This evergreen shrub is native to Mediterranean Europe and northern Africa. The white, fragrant flowers open from pink buds and are held in clusters. Dark, glossy blue berries follow the blooms and are popular with birds. In fact, it is believed that the sheen on the berries, which is almost metallic, is due to a fat structure around the berries that is designed to attract the birds whose droppings disperse the seeds.

WITCH HAZEL
Hamamelis virginiana

This flowering shrub, originally from eastern North America, was first thought to be medicinally beneficial due to the shape of its flowers. Indigenous American peoples have long used witch hazel to treat sore or inflamed skin, a practice which was picked up by European settlers. European physicians considered the curling flowers to be a representative of the snake coiling around the staff of Asclepius, the Greek god of medicine. The plant is still widely used today.

The curling snake that witch hazel flowers were thought to resemble, can be seen on this marble Roman statue of Asclepius, 2nd century CE, after a Greek original of the 5th century CE.

HELLEBORE
Helleborus orientalis

The markings seen around the centre of the hellebore flowers are known as freckles.

Hellebores are known as the Christmas rose as they start flowering around this time of year; despite having no relation to wild *Rosa* species, the flowers do bear a passing resemblance. The oriental types, which are originally from Greece, Turkey and surrounding areas, hybridize and self-seed easily in the garden. The result of this has been further flowers in a wide range of colours. A popular way to display the flowers, many of which naturally hang down, is to cut off the heads and place them in a shallow dish of water. They will float and last a few days as an attractive floral display.

COMMON PRIMROSE
Primula vulgaris

The yellow flower seen here is *Primrose Nymph* illustrated by Walter Crane (1889).

Native to northwest Africa, western and southern Europe, as well as parts of southwest Asia, primroses prefer to grow in woods, grasslands and at the base of hedgerows. The flower heads can sometimes be found scattered on the ground; this is due to greenfinches pulling off the flowers to eat the nectaries and ovaries. According to Irish folklore, keeping primroses in your doorway protects your home from fairies.

SWEET BOX

Sarcococca hookeriana var. *digyna*

Also known as sweet box, this low-growing, evergreen shrub produces highly scented white flowers in winter, followed by black berries. Native to China, Afghanistan, Nepal and Bhutan, *digyna* is more commonly grown than the species plant *Sarcococca hookeriana* itself, as the stems and leaves are more slender. The genus name *Sarcococca* comes from the Greek meaning 'fleshy berries'.

Sweet box is commonly used as hedging, providing fragrant flowers this month, here lining the path at Anglesey Abbey, Cambridgeshire.

DAISY
Bellis perennis

The common name of this plant comes from 'day's eye', as the flowers open each morning and close each night. Each daisy comprises around 30–60 petals and the random number lends itself to a game of French origin, *Effeuiller la marguerite* ('He/she loves me, he/she loves me not'). In the game, these phrases are spoken as the petals are pulled off one by one, the final petal supposedly revealing the truth. While the English rhyme alternates only between 'loves me and 'loves me not', in the French version the petals also reveal how much the person is loved: 'a little', 'a lot', 'to madness' or 'not at all'.

WHITE GINGER LILY
Hedychium coronarium

In the tropical forests of Asia where the white ginger lily is found, it can be seen flowering around this time of year.

Native to the forests of India, Bangladesh, Nepal, Bhutan and China, white ginger lily is found growing on the forest floor. It is cultivated around the world for its ornamental value and use in perfume production; its scent is said to be reminiscent of jasmine. It is the national flower of Cuba and during Spanish colonial times it was worn by women who used the intricate shape of the flowers to conceal and transport secret messages to help in the fight for independence.

DARWIN'S SLIPPER
Calceolaria uniflora

The quirky flowers of Darwin's slipper can be seen blooming up in the rocky hills of the Andes and Patagonia this month.

These unusual flowers are thought to resemble little shoes, from which the word *Calceolaria* is derived. This particular species is famous for being spotted by Charles Darwin while he was exploring South America. However, it was actually first collected by French botanist Philibert Commerson in 1767. It is found growing along the rocky peaks of the Andes and Patagonia and is pollinated by birds, namely seedsnipes, who peck at the white part of the flower that is high in sugar. Pollen transfers to the heads of the birds as they do this, and then between plants as the birds move around to feed.

PINEAPPLE LILY
Eucomis comosa

In its native land of South Africa, the pineapple lily flowers throughout the summer.

Native to South Africa, the flowers of the pineapple lily grow in dense clusters, forming cylindrical spikes. They are topped with green, leaf-like bracts, creating the look of a pineapple. The Latin name *Eucomis* is derived from the Greek words 'eu' meaning 'good' and 'kome' meaning 'hair', relating to the tufts of leaves crowning the plant. The flowers are sweet-smelling, and once they are pollinated by wasps and bees they close, turning a dark brown and later revealing seeds.

SUMMER HYACINTH
Ornithogalum candicans

Previously known as *Galtonia candicans*, the elegant summer snowdrop produces spikes of bell-shaped flowers in South Africa's summer months.

The summer hyacinth produces fragrant, bell-shaped flowers on spikes, which can grow up to 1.5m (5ft) in height. The species name *candicans* means 'becoming pure white', referring to the flowers. This regal plant originates from South Africa; it was first thought to be a type of hyacinth, but was later renamed. Popular with bees and butterflies, it grows from bulbs which are also grown in gardens outside South Africa. It resembles a giant snowdrop and is popular as a garden plant.

GLORY LILY
Gloriosa superba

Flowering during December in South Africa, the glory lily produces elegant and beautiful but poisonous blooms.

Also known as the flame lily, this plant is native to southeastern Africa and South Asia, and is the national flower of Zimbabwe. A diamond brooch in the shape of the glory lily was presented to Princess Elizabeth (now Queen Elizabeth II) in 1947 on her visit to the country that was then known as Southern Rhodesia. It is a popular flower with florists, but despite its beauty, it is poisonous and is sufficiently toxic to be fatal to humans and animals if consumed.

AFRICAN DAISY
Dimorphotheca jucunda

In the mountains of South Africa, the African daisy blooms throughout the spring and summer months of the southern hemisphere.

Growing from underground stems, which are believed to help the plants survive fires and cold winters, this pretty perennial is found in the mountains of South Africa. The species name *jucunda* comes from the Latin meaning 'pleasing', 'delightful' or 'lovely'. The flowers are large with pale magenta petals and are popular with butterflies. This is the parent plant of some of the many African daisies we see in gardens and bouquets today.

BLUE MARGUERITE DAISY
Felicia amelloides

The blue marguerite daisy produces many blue flowers that sit high above the plant's leaves this month in South Africa.

This daisy is found along the southern coasts of South Africa, growing on sand dunes, exposed hillsides and basalt cliffs. Its ability to withstand dry and windy conditions means it has been planted on sand dunes to help stabilize the soil. It is also commonly grown in gardens, its bright blue flowers making it popular with gardeners. After flowering, round fluffy seed heads are produced; each seed has an appendage that will act as a tiny parachute, helping it to disperse further away from the parent plant.

CAMELLIA 'CRIMSON KING'
Camellia sasanqua 'Crimson King'

Camellia 'Crimson King' is a very valuable shrub for gardens, providing big showy blooms in December.

This plant needs a sheltered position if it is to be grown outside in temperate regions. However, it provides glorious colour in the depths of winter, even December, if properly protected. The species name *sasanqua* is derived from the Japanese name for this plant *sazanka* and the 'Crimson King' cultivar is a popular hybrid. Tea oil is produced from this species, which is used in China by those working with lacquer to remove varnish from the skin.

MOTH ORCHID
Phalaenopsis amabilis

The close-up of this moth orchid reveals the intricate markings that attract pollinators.

Moth orchids are the easiest of the orchids to keep flowering, which is why they are the most popular and recommended for orchid enthusiasts that are just starting out. This species is one of the first parent plants used to breed the many cultivars available today, with the first hybrid created in England in 1875, at the Veitch & Sons nurseries in Devon. The moth orchid is native to the East Indies and Australia, and, like other orchids, can live for many years. In the wild these orchids grow up trees or other vegetation, gaining nutrients and water from the air or plant matter around them.

PINK POWDERPUFF
Calliandra haematocephala

Each flower of the pink powderpuff tree is made up of many striking bright pink stamens.

Native to Bolivia, this small ornamental tree is very popular in places like Florida, where it enjoys the climate. Used in traditional medicine for its healing properties, research has shown that it contains chemical compounds that could help in the treatment of cancer. The large, fluffy blooms are fragrant and the main reason for its popularity; however, it is the stamens rather than the petals that give the flowers their eye-catching appearance.

RED HOT POKERS
Kniphofia uvaria

The fiery spikes of red hot pokers bloom throughout December in their native South Africa.

Originating from the Cape Province of South Africa, this plant is now grown in gardens around the world for its colourful, exotic flowers. As a garden escapee it has become invasive and is now considered a weed, causing damage to ecosystems where it grows in big clumps, such as in parts of southeastern Australia. With narrow leaves shaped like those of a lily, the flower heads grow as tall as 1.5m (5ft), and can create an impressive sight.

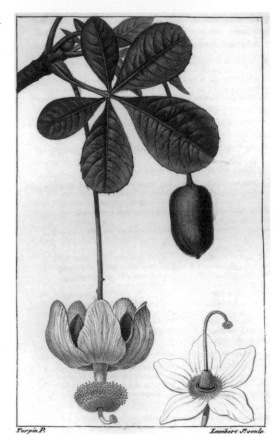

Turpin P.   *Lambert Sc sculp*

BAOBAB
Adansonia digitata

Hand-coloured
stipple
copperplate
engraving of
the flower of a
baobab after a
drawing by Pierre
Jean-Francois
Turpin in *La Flore
Médicale*, 1830.

Found across the savannahs of sub-Saharan Africa, the baobab is known to live for up to 2,000 years; its flowers, however, last for just one day. The blooms are white and pendulous with a sour scent; they open at night and are pollinated by insects. Each tree can grow up to 25m (80 ft) high and is revered as a symbolic home to the spirits of dead ancestors. The flowers are followed by velvety fruit, which is high in vitamin C and is made into a refreshing drink.

HOLLY
Ilex aquifolium

Following the flowers in spring and early summer, the recognisable red berries of holly form this month.

Native to western and southern Europe, northwest Africa and western Asia, holly is one of the few native evergreen trees in the UK. It is recognized by its bright red berries, which are produced on the female plant following the white flowers (both male and female plants produce flowers). The male flowers have a small dark dome in the centre with an opening resembling the head of a Phillips screwdriver. Before it was associated with Christmas, the Druids, Celts and Romans all celebrated this plant, believing holly's ability to remain green throughout winter was magical and assured the return of spring.

CLOVE
Syzygium aromaticum

Harvesting the buds this time of year in Indonesia before the flowers develop provides the spice that is used in cooking.

Native to the Maluku Islands of eastern Indonesia, the aromatic spice we use in cooking is the dried flower buds. Clove oil contains an antiseptic and anaesthetic oil called eugenol, which is why it is used in traditional medicines. Oranges studded with cloves are used as an insect repellent in the Maluku Islands, and elsewhere in the world these have been used as pomanders, historically to mask unpleasant smells, but more recently at Christmas time to create a festive atmosphere. The plant flowers at this time of year in East Africa.

CLEMATIS 'JINGLE BELLS'
Clematis cirrhosa 'Jingle Bells'

Clematis 'Jingle Bells' produces a prolific number of flowers this month, brightening up winter gardens.

Believed to be the hardiest winter-flowering clematis, coping well with cold weather in temperate regions, this flower, with its suitably festive name, not only flowers in December but will continue through to March. It is a popular garden climber, known for its prolific, creamy-white, nodding blooms and vigorous growth. If grown in a sunny position it will develop a lemony scent.

CHRISTMAS CACTUS
Schlumbergera truncata

This flower is known as the Christmas cactus due to its timely flowers. Later in the year, a very similar plant, the *Schlumbergera gaertneri* which flowers around April, is likewise known as the Easter cactus.

This plant flowers from late November to late January in the Northern Hemisphere, providing a wonderful display over the festive period. Unlike desert-dwelling cacti, this succulent prefers semi-shade and makes an easy-to-care-for house plant where it isn't warm enough for it to be grown outside. It originates from the coastal mountains of southeast Brazil, where the air is warm with high humidity.

SNOWDROP 'THREE SHIPS'
Galanthus plicatus 'Three Ships'

Snowdrops 'Three Ships' flowers unusually early, which is why it is possible to see this in flower on Christmas Day.

Snowdrops were introduced to the UK as far back as the late 16th century. Since then they have been cultivated and become naturalized. This variety was discovered in 1984 growing under an ancient cork oak at Henham Park, Suffolk, by nurseryman John Morley. The colony was noted for its unusually early flowering, and is typically in flower by Christmas Day. The name Morley chose cleverly references a popular Christmas carol that originates from the 17th century.

I saw three ships come sailing in
On Christmas Day, on Christmas Day;
I saw three ships come sailing in
On Christmas Day in the morning.

ANON

POINSETTIA
Euphorbia pulcherrima

In the northern hemisphere poinsettia are grown inside glasshouses and therefore ready in time for Christmas despite the cold weather.

Known for its red and green foliage, this plant from Mexico and Central America has become a festive plant, commonly seen at Christmas. The association with this holiday began in Mexico in the 16th century, when, according to legend, a girl was too poor to buy a gift for the celebration of Christ's birth. An angel inspired her to instead gather roadside weeds and place them at the altar of the church, whereupon red blooms came up from the weeds and the plants became poinsettias.

The bulbs of amaryllis 'Carmen' are also grown indoors in order to bloom for Christmas in cooler climates, with each huge flowers lasting several weeks.

AMARYLLIS 'CARMEN'

Hippeastrum 'Carmen'

Now a popular plant grown around Christmas, a hybrid of amaryllis was first bred for indoor cultivation in 1799 by English watchmaker, Arthur Johnson. While his own greenhouse was destroyed in a fire, he had fortunately shared his plants with the Liverpool Botanic Garden, and by the mid-19th century this amaryllis was even being bred in the USA. Native to parts of Argentina, Mexico and the Caribbean, the genus name is derived from the Ancient Greek, meaning a 'knight's star', while the common name comes from the Greek 'to sparkle'.

KAHILI GINGER
Hedychium gardnerianum

Kahili ginger produces very sweetly scented blooms, flowering until December in the Himalayas.

This member of the ginger family is native to the Himalayas. Bright green leaves grow up the tall stems, and the plant produces very fragrant yellow and red flowers, which smell sweet and faintly of ginger. While it prefers a tropical climate, it has been known to survive an occasional frost, meaning it can be grown in some temperate gardens around the world. Recent research has suggested that the plant may be able to fight against human small-cell lung cancer as it contains the plant compound villosin.

WINTER HELIOTROPE
Petasites fragrans

The blooms of winter heliotrope can be spotted in December; the clusters of flowers with very short petals are vanilla scented.

Although originally from North Africa, this plant is now common in temperate regions of the Northern Hemisphere, including the UK. It can be found on roadsides, rough ground or by streams. Scented strongly of vanilla, forming patches of heart-shaped leaves, this flower is a valuable winter food source for bees.

WINTER DAPHNE
Daphne odora

This evergreen shrub produces flowers at the end of the stems, making them easy to spot in the dark winter months; the cultivar shown here is *Daphne odora* 'Marianni'.

The species name *odora* refers to the strong fragrance produced by this plant. Originally from China, it is grown in temperate gardens of the Northern Hemisphere for winter interest. The foliage is glossy, and while it rarely produces fruit, the red berries can appear after flowering. The fragrance produced by the clusters of flowers is sweet and spicy, suitably festive for this time of year.

CALLA LILY
Zantedeschia aethiopica

The luxurious exotic calla lily blooms seen close-up in *Two Calla Lilies on Pink*, by Georgia O'Keeffe, 1928.

Much of the abstract work of Georgia O'Keeffe has been interpreted as strongly sexual despite her denying such intentions. Her painting *Two Calla Lilies on Pink* (1928) shows off the beautiful form of these exotic flowers. Native to parts of Africa, they are originally believed to symbolize purity and innocence. However, the elegant, sculptural blooms have continued to be used in bouquets and art worldwide, often suggesting opulence and luxury.

INDEX

abelia, glossy 344
Abeliophyllum distichum 56
Abraham-Isaac-Jacob 46
Acacia
 A. *dealbata* 78
 A. *pycnantha* 330
Acer
 A. *palmatum* 114
 A. *pseudoplatanus* 117
Achillea millefolium 168
Achilles 168
Acis autumnalis 335
aconite 210
 winter 50
Aconitum napellus 210
Adansonia digitata 366
Adonis annua 247
Aesculus hippocastanum 140
African daisy 360
African lily 25
Agapanthus africanus 25
Agrostemma githago 275
air plant 84
Alcea rosea 45
Alchemilla mollis 171
Algerian iris 28
alkanet, green 113
Allium
 A. *siculum* 158
 A. *thunbergii* 'Ozawa' 345
 A. *ursinum* 180
almond 54
aloha, dark sky-blue 332
Alstroemeria aurea 197
Althaea officinalis 246
alyssum, sweet 111
Amaranthus cruentus 208
amaryllis 'Carmen' 373
Amazon moonflower 136
Amelanchier lamarckii 122
Ammi majus 184
Amorphophallus titanum 225
Anagallis arvensis 228
anchor plant 343
Anemone
 A. *blanda* 68
 A. *coronaria* 99
anemone of Passiontide 97
angelica 174
Angelica archangelica 174
angel's trumpet 286
Anigozanthos rufus 324

aniseed 190
Anne of Denmark **212**
Antirrhinum majus 254
aphrodisiacs 25
Aphrodite 95
Apollo 89
apple 108
apple of Peru 301
Aquilegia vulgaris 121
arbor Judea 75
Arbutus unedo 311
Arctium lappa 248
Arcus, Willibald 130
Argyranthemum frutescens 249
Armeria maritima 197
arnica 235
Arnica montana 235
Artemisia vulgaris 246
Arum maculatum 130
Asclepius 349
ashwagandha 334
Asian watermeal 274
aspen, quaking 58
Aster
 A. *amellus* 283
 A. *tataricus* 332
astilbe 268
Astilbe rubra 268
Atropa belladonna 256
Audrey II 100
Augustus, Emperor 256
Augustus Frederick, Prince 221
autumn crocus 296
autumn snowflake 335
azalea rhododendron 134

baby's breath 250
bacopa 24
bamboo, common 119
Bambusa vulgaris 119
banana 23
banana water lily 245
Banksia baxteri 83
baobab 366
Baptisia australis 127
Barclay, George **62**
Barringtonia asiatica 222
basil, holy 21
beach cabbage 230
Beaton, Cecil **330**
bee orchid 201
beech drops 241
begonia 186
Begonia gracilis 186

bell heather 43
bell vine, purple 213
belladonna 256
bellflower, peach-leaved 188
Bellis perennis 354
bells of Ireland 278
Bergenia crassifolia 98
Berlese, Lorenzo 51
Bessera elegans 193
big blue lilyturf 342
bilberry 94
bindweed, blue rock 328
bird of paradise 316
bird's-foot trefoil 263
bird's nest banksia 83
bishop's flower 184
black-eyed Susan vine 325
blackthorn 312
bleeding heart 57
blue marguerite daisy 361
blue rock bindweed 328
blue throatwort 141
blue tweedia 336
blue wild indigo 127
bluebell, English 104
blueblossom 135
bombweed 214
borage 272
 early-flowering 46
Borago officinalis 272
Bouvardia ternifolia 323
Bowden lily 17
Bowie, James 294
Bowles, Edward Augustus 61
box fruit 222
Brugmansia arborea 286
Buddhism 196
Buddleja davidii 315
burdock, greater 248
Burnett, Mary Ann **210**
buttercup
 meadow 167
 Persian 151
butterfly bush 303, 315
Byzantine gladiolus 144

Calceolaria uniflora 356
Calendula officinalis 295
Calibrachoa parviflora 131
California poppy 202
calla lily 377
Calliandra haematocephala 364
Calluna vulgaris 320

camas, common 162
Camassia quamash 162
Camellia
 C. *japonica* 51
 C. *sasanqua* 'Crimson King' 362
 C. *sinensis* 15
camellia, common 51
Campanula
 C. *persicifolia* 188
 C. *rotundifolia* 260
candle larkspur 206
cape primrose 294
Cardamine
 C. *pratensis* 153
 C. *trifolia* 73
carnation 289
carrion flower 33
carrot, wild 212
Catalpa speciosa 192
Catesby, Mark 253
cathedral bells 211
catnip 290
Ceanothus thyrsiflorus 135
Centaurea cyanus 169
Cercis siliquastrum 75
Cerinthe major 187
Chaenomeles speciosa 91
Chaenostoma cordatum 24
Chamaenerion angustifolium 214
Chamelaucium uncinatum 325
chamomile, German 166
Charlemagne 223
Charles I 220
chaste tree 216
cherry 69
cherry, Cornelian 27
chickweed, common 41
chicory 229
Chilean lantern tree 19
Chimonanthus praecox 29
Chinese fringe flower 65
Chinese hibiscus 310
Chinese meadow-rue 251
Chinese quince 91
Chionanthus virginicus 139
Christmas cactus 370
chrysanthemum 339
Chrysanthemum spp. 339
Cichorium intybus 229
cigar tree 192
Cirsium vulgare 221
Citrus × aurantium 93
Clark, William 234

clematis
  evergreen 31
  'Jingle Bells' 369
  'Winter Beauty' 13
Clematis
  C. cirrhosa 31
  C. cirrhosa 'Jingle
    Bells' 369
  C. urophylla 'Winter
    Beauty' 13
clove 368
clover, white 86
Cobaea scandens 211
Cobo, Bernabé 211
Colchicum autumnale 296
Colletia paradoxa 343
coltsfoot 92
columbine 121
Commerson, Philibert
  356
coneflower 269
  purple 298
conkers 140
Convallaria majalis 124
Convolvulus sabatius 328
Cooke, George 25
coral drops 193
corncockle 275
cornelian cherry 27
cornflower 169
Cornus
  C. florida 105
  C. mas 27
  C. sanguinea 185
Coronilla valentina subsp.
  glauca 40
corpse flower 195
Corydalis nobilis 67
Corylus avellana 52
Cosmos bipinnatus 178
coughwort 92
courtship and matrimony
  224
Coventry, Lord 29
cowslip 123
Crambe
  C. cordifolia 182
  C. maritima 200
cranesbill, meadow 257
crape myrtle 215
Crataegus monogyna 155
crimson flag lily 88
crimson scabious 255
Crinodendron hookerianum
  19
Crocosmia aurea 22
Crocus sativus 318
crown imperial 96
cuckoo flower 153
  three-leaved 73
Culpeper, Nicholas 172
cup and saucer vine 211
Cupani, Franciscus 194

Cupid 188
curry plant 261
Cyclamen hederifolium 304
Cydonia oblonga 95
Cypripedium calceolus 179

Dactylorhiza fuchsii 198
daffodil
  'February Gold' 59
  'Rijnveld's Early
    Sensation' 38
  wild 72
Dahlia
  'Bishop of Llandaff'
    277
  'Blyton Softer Gleam'
    8
daisy 354
dandelion 322
Daphne
  D. bholua 39
  D. mezereum 53
  D. odora 376
dark sky-blue aloha 332
Darwin, Charles 211, 356
Darwin's slipper 356
Daucus carota 212
Davidia involucrata 116
de Belder, Robert 26
De Graaff Brothers 59
de Launay, Mordant 94
de Pannemaeker, J. 202
de Passe, Simon 212
deadly nightshade 256
Delphinium elatum 206
di Giorgio Martini,
  Francesco **95**
Dianthus
  D. barbatus 207
  D. caryophyllus 289
Diascia mollis 48
Digitalis purpurea 170
Dimorphotheca jucunda
  360
Dionysus 319
Dior, Christian 124
Dipsacus fullonum 265
dog rose 159
dog's tooth violet 82
dogwood
  common 185
  flowering 105
Douglas, David 66
dove tree 116
Downey, W. & D. 289
Drake, Sarah **62**
Du Fu 134
dwarf crown imperial 109

Easter lily 118
Easter tree 76
Echinacea purpurea 298
Edgeworthia tomentosa 62

eggs and bacon 263
Elaeagnus × submacrophylla
  341
Elaeagnus angustifolia 160
elderflower 154
elephant's ears 98
Elizabeth I 282
Elizabeth II 330, 359
empress tree 105
English bluebell 104
The English Flower Garden
  (Robinson) 68, 331
English oak 115
Epifagus virginiana 241
Epigaea repens 132
Eranthis hyemalis 50
Erica cinerea 43
Erodium cicutarium 235
Eryngium variifolium 264
Erysimum
  E. bicolor 79
  E. 'Bowles's Mauve' 61
Erythronium dens-canis 82
Eschscholzia californica 202
Eucomis comosa 357
Euphorbia
  E. helioscopia 297
  E. pulcherrima 372
Euphrasia officinalis 172
European horse chestnut
  140
European Michaelmas
  daisy 283
Eustoma grandiflorum 273
evening primrose 234
evergreen clematis 31
eyebright 172

fairies 153, 155, 159, 170,
  260, 320, 351
falling stars 22
false goat's beard 268
Felicia amelloides 361
feverfew 233
fiddle neck 237
field scabious 299
Filipendula ulmaria 224
fingers and thumbs 150
firecracker bush 323
Fitch, Walter **22**
flaxseed 242
flower anatomy **11**
flowering dogwood 105
folklore *see also*
  mythology;
  superstitions
  blue wild indigo 127
  carrion flower 33
  chicory 229
  common dogwood
    185
  common lungwort 49
  common stock 202

crown imperial 96
Easter lily 118
guelder rose 145
harebell 260
hyacinth 89
hydrangea 165
lady's bedstraw 262
meadow buttercup
  167
pasqueflower 97
primrose 351
rosemary 163
rowan 156
snapdragon 254
Star of Bethlehem 126
verbena 311
weeping forsythia 76
wild strawberry 157
Ford, H. J. **175**
forget-me-not 288
Forrest, George 60
forsythia
  weeping 76
  white 56
Forsythia suspensa 76
Fortune, Robert 24
Fortuny, Mariá 148
foxglove 170
foxglove tree 105
Fragaria vesca 157
frangipani 103
freesia 162
Freesia refracta 162
French lavender 252
fringe cups 152
Fritillaria
  F. imperialis 96
  F. meleagris 102
  F. raddeana 109
Fuchsia
  F. hatschbachii 281
  F. triphylla 209

Galanthus
  G. nivalis 12
  G. plicatus 'Three
    Ships' 371
Galium verum 262
garden cosmos 178
garden lobelia 347
gardenia, common 77
Gardenia jasminoides 77
garlic, wild 180
Garrya elliptica 66
Gaura lindheimeri 308
Gentian sage 331
Geraldton wax flower 325
Geranium pratense 257
geraniums 14
Gerard, John 240
gerbera daisy 337
Gerbera jamesonii 337
German chamomile 166

Geum
  G. rivale 259
  G. urbanum 258
ghost plant 243
ghost tree 116
Gladiolus 10
  G. communis subsp.
    byzantinus 144
glaucous scorpion-vetch
  40
Gloriosa superba 359
glory lily 359
glory-of-the-snow 74
glossy abelia 344
goat willow 85
golden-beard penstemon
  218
golden wattle 330
gooseneck loosestrife 217
gorse 138
grape hyacinth 128
great blue lobelia 280
green alkanet 113
guelder rose 145
Gypsophila paniculata 250
gypsy fern 150

Halesia carolina 112
Hamamelis
  H × intermedia 'Jelena'
    26
  H. virginiana 349
Hamlet (Shakespeare) 163
handkerchief tree 116
harebell 260
Harris, Moses 333
Hatschbach's fuchsia 281
hawthorn 155
hazel, common 52
heart-leaf bergenia 98
heather 320
  bell heather 43
Hedera helix 319
Hedychium
  H. coronarium 355
  H. gardnerianum 374
Helenium 'Ribinzwerg' 9
Helianthus annuus 220
Helichrysum italicum 261
Heliconia rostrata 307
heliotrope 87
Heliotropium arborescens
  87
hellebore 350
  stinking 37
Helleborus
  H. foetidus 37
  H. orientalis 350
herb bennet 258
The Herball or Generall
  Historie of Plantes
  (Gerard) 240

Hesperantha coccinea 88
Hey, Rebecca 156
Hibiscus rosa-sinensis 310
Himalayan balsam 215
Himalayan blue poppy
  173
Hinduism 21, 196, 310
Hippeastrum 'Carmen' 373
Holboellia coriacea 120
Holiday, Billie 77
holly 367
hollyhock 45
holy basil 21
Homer 27
honeysuckle
  purpus 36
  winter-flowering 16
honeywort 187
Hood Fitch, Walter 181
Hopkins, Arthur 123
horse chestnut, European
  140
houseleek, common 223
Humphreys, Henry
  Noel 67
hyacinth, common 89
Hyacinthoides non-scripta
  104
Hyacinthus 89
Hyacinthus orientalis 89
hydrangea 165
Hydrangea macrophylla 165
Hypericum perforatum 314
hyssop 240
Hyssopus officinalis 240

'I Wandered Lonely As a
  Cloud' (Wordsworth)
  72
Ilex aquifolium 367
Impatiens glandulifera 215
'In Flanders Field'
  (McCrae) 326
Ipomoea purpurea 305
Iris
  I. damascena 80
  I. reticulata 58
  I. unguicularis 28
irises
  Algerian 28
  crimson flag lily 88
  early bulbous 58
  species 80
ironwood, Persian 39
ivy, common 319
ivy-leaved cyclamen 304

Jacobaea vulgaris 333
jade vine 129
James V of Scotland 221
Japanese andromeda 68
Japanese apricot 34

Japanese maple 114
Japanese onion 345
Japanese pink pussy
  willow 64
Japanese snowbell 137
jasmine, common 226
jasmine, winter 24
Jasminum
  J. nudiflorum 24
  J. officinale 226
Jefferson, Thomas 300
Jekyll, Gertrude 68
Johnson, Arthur 373
Judaism 238
Judas tree 75
Jung, Johann Jakob 51

kahili ginger 374
Kali 310
kangaroo paw 324
Karwinsky, Wilhelm 193
king protea 285
Kingdon-Ward, Frank 173
kiss-me-over-the-garden-
  gate 300
Knautia
  K. arvensis 299
  K. macedonica 255
Kniphofia uvaria 365
Krishna 21

Laburnum anagyroides 147
laburnum, common 147
lacy phacelia 237
lady-in-a-bath 57
lady's bedstraw 262
lady's eardrop 209
lady's mantle 171
lady's slipper orchid 179
lady's smock 153
Lagerstroemia indica 215
Lamium album 329
Lamprocapnos spectabilis 57
Lantana camara 18
lantana, common 18
lantern tree, Chilean 19
larkspur, candle 206
Lathyrus odoratus 194
laurustinus 348
Lavandula stoechas 252
lavender, French 252
Laxmann, Erik 67
lemon balm 306
Lenihan, Eddie 155
Leonardo da Vinci 114
Leptospermum scoparium
  32
Leucojum aestivum 110
lilac, common 148
lilies
  African lily / lily of
    the Nile 25

Bowden lily 17
calla lily 377
Easter lily 118
glory lily 359
pineapple lily 357
white ginger lily 355
Lilium longiflorum 118
lily of the valley 124
lily-of-the-valley shrub
  68
lilyturf, big blue 342
Limonium vulgare 232
Linaria vulgaris 317
Linden, Jean 202
Linnaea × grandiflora 344
Linnaeus, Carl 8, 10, 67
linseed 242
Linum usitatissimum 242
Liriodendron tulipifera 161
Liriope muscari 342
lisianthus 273
Little Shop of Horrors (1986
  film) 100
Livia Drusilla 256
Lobelia
  L. erinus 347
  L. siphilitica 280
lobster claw 307
Lobularia maritima 111
Lonicera
  L. × purpusii 36
  L. fragrantissima 16
loosestrife, gooseneck 217
lords-and-ladies 130
Loropetalum chinense 65
lotus 196
Lotus corniculatus 263
Loudon, Jane 67, 331
Loudon, John 29
Louise of Prussia 169
love-in-a-mist 231
lungwort, common 49
lupin, narrow-leaved
  204–5
Lupinus angustifolius
  204–5
Lysimachia clethroides 217
Lysimachus of Thrace 217

Magnolia grandiflora 253
Mahonia × media 'Winter
  Sun' 30
Malus domestica 108
Malva moschata 229
Manuka 32
maple, Japanese 114
marguerite 249
marigold
  Mexican 189
  pot 295
Marsdenia floribunda 292
marsh mallow 246

Mary, Queen of Scots 221
Matheus, Jean 27
Matricaria chamomilla 166
Matthiola incana 202
mayflower 132
McCrae, John 326
meadow buttercup 167
meadow cranesbill 257
meadow saffron 296
meadowsweet 224
Meconopsis betonicifolia 173
Medinilla magnifica 181
Melissa officianalis 306
Metamorphoses (Ovid) 27
Mexican ivy 211
Mexican marigold 189
Mexican sunflower 338
mezereon 53
Michaelmas daisy,
  European 283
million bells 131
mimosa 78
mock orange 133
Moluccella laevis 278
Monet, Claude 147, 244–5
monk's hood 210
Monotropa uniflora 243
montbretia 22
Morley, John 371
morning glory, common
  305
Moroccan sea holly 264
Morris, Richard **234**
moth orchid 363
mountain camellia 219
mugwort 246
Müller, Walther 115,
  174, 322
Musa acuminata 23
Muscari armeniacum 128
musk mallow 229
Mussin-Pushkin, Apollos
  Apollosovich 81
Myosotis scorpioides 288
mythology
  common houseleek
    223
  Greek 17, 27, 38, 89,
    95, 148, 168, 188, 217,
    285, 319, 342, 349
  Hindu 21
  Mexican 372
  Pied Piper of Hamlin
    175

naked ladies 296
Narcissus 38
Narcissus
  'February Gold' 59
  N. pseudonarcissus 72
  'Rijnveld's Early
    Sensation' 38

narra 42
nasturtium 47
national emblems 6, 42,
  71, 221, 242, 271, 310,
  324, 359
Nelumbo nucifera 196
Nemesia caerulea 332
Nepalese paper plant 39
Nepeta cataria 290
Nereids 17
Nerine bowdenii 17
nettle 287
Nicandra physalodes 301
Nicot de Villemain, Jean
  284
Nicotiana rustica 284
nigella 231
Nigella damascena 231
night-scented
  pelargonium 14
night-scented phlox 20
nightshade, deadly 256
nomenclature 8, 10–11
Nymphaea mexicana 245

oak, English 115
Ocimum tenuiflorum 21
Odyssey (Homer) 27
Oenothera biennis 234
O'Keeffe, Georgia **377**
oleaster 341
Ophrys apifera 201
orange, Seville 93
orange witch hazel 26
orchids
  bee 201
  common spotted 198
  lady's slipper 179
  moth orchid 363
  vanilla 282
  waling-waling 271
Oregon grape 'winter
  sun' 30
Ornithogalum
  O. candicans 358
  O. umbellatum 126
Ovid 27
Oxypetalum coeruleum 336

Paeonia lactiflora 149
Pan 148
Papaver
  P. cambrica 227
  P. rhoeas 326
paperbush 62
Parrotia persica 39
parsley 238
pasqueflower 97
Passiflora edulis 276
passionflower 276
Paulownia tomentosa 105
peace lily 183

peach-leaved bellflower
  188
Penstemon barbatus 218
Pentaglottis sempervirens
  113
peony 149
Perlagonium triste 14
Persian buttercup 151
Persian ironwood 39
Persicaria orientalis 300
Peruvian lily 197
Petasites fragrans 375
Petre, Robert James 51
Petroselinum crispum 238
petunia, wild 302
Phacelia tanacetifolia 237
Phalaenopsis amabilis 363
pheasant's eye 247
Philadelphus coronarius 133
phlox 270
  night-scented 20
Phlox paniculata 270
Pied Piper of Hamlin 175
Pieris japonica 68
pig squeak 98
Pimpinella anisum 190
pineapple lily 357
pink powderpuff 364
pitcher plant, pale 191
plant names 8, 10–11
Plato 216
Pliny the Elder 92, 261,
  311
plum 34
  common 63
Plumeria rubra 103
Plumier, Charles 209
Pochin, Henry Davis 60
poinsettia 372
polyanthus 'Gold Laced'
  90
Polygonatum multiflorum
  177
Ponting, Herbert **339**
poorhouse flowers 291
poppers 150
poppy
  California 202
  common 326
  Himalayan blue 173
  Welsh 227
poppy anemone 99
Populus tremula 58
pot marigold 295
potato vine 239
primrose, common 351
Primula
  'Gold Laced' 90
  P. veris 123
  P. vulgaris 351
Protea cynaroides 285
Proteus 285

Prunella vulgaris 309
Prunus
  P. × yedoensis 'Somei-
    yoshino' 69
  P. amygdalus 54
  P. domestica 63
  P. mume 34
  P. spinosa 312
  P. subhirtella
    'Autumnalis Rosea'
    346
Pseudofumaria lutea 150
Pterocarpus indicus 42
Puddle, Charles 60
Pulmonaria officinalis 49
Pulsatilla vulgaris 97
purple bell vine 213
purple coneflower 298
purpus honeysuckle 36
Puschkinia scilloides 81
pussy willow 85
  Japanese pink 64

quaking aspen 58
Queen Anne's lace 212
Queen of Philippine 271
Quercus robur 115
quince 95
  Chinese 91

Rafflesia arnoldii 195
ragwort 333
Ranunculus
  R. asiaticus 151
  R. repens 167
red amaranth 208
red hot pokers 365
Redouté, Pierre-Joseph
  140, 164, 286
Reeves, John 142
Rhinanthus minor 266–7
Rhodochiton atrosanguineus
  213
Rhododendron
  R. indicum 134
  R. ponticum 107
rhododendron, common
  107
Robinson, William 68,
  331
Rosa
  R. × centifolia 164
  R. canina 159
rose campion 279
rose grape 181
rose of a hundred petals
  164
rosebay willowherb 214
rosemary 163
Rotheca myricoides 303
rowan 156
Rudbeckia hirta 269

Ruelle, Jean 302
Ruellia humilis 302
Rumex acetosa 176
Russian olive 160

saffron crocus 318
sage, yellow 18
Saint Patrick 86
sakura 69
Salix
    S. *caprea* 85
    S. *gracilistyla* 'Mount
        Aso' 64
Salvia
    S. *patens* 331
    S. *rosmarinus* 163
Sambucus nigra 154
Sarcococca hookeriana var.
    digyna 353
Sarracenia alata 191
sausage vine 120
Scabiosa columbaria 293
scabious
    crimson 255
    field 299
    small 293
Scaevola taccada 230
scarlet pimpernel 228
Schizanthus pinnatus 144
Schlumbergera truncata
    370
Scilla forbesii 74
scorpion-vetch, glaucous
    40
sea holly, Moroccan 264
sea kale 200
    flowering 182
sea lavender 232
sea poison tree 222
sea thrift 197
Selenicereus wittii 136
selfheal 309
Sempervivum tectorum 223
serviceberry 122
Seville orange 93
Shakespeare, William
    134, 163, 207
shamrock 86
Shennong 15
shoo-fly plant 301
Siberian corydalis 67
Sicilian honey garlic 158
Silene coronaria 279
silk tassel bush 66
silverberry 341
snapdragon 254
snowdrop
    common 12
    'Three Ships' 371
snowdrop tree 112
Solanum laxum 239
Solomon's seal 177

Sorbus aucuparia 156
sorrel 176
Spanish bluebell 104
Spathiphyllum wallisii 183
spear thistle 221
Species Plantarum
    (Linnnaeus) 8
squill, striped 81
St John's wort 314
stachyurus, early 63
Stachyurus praecox 63
Stapelia gigantea 33
star jasmine 91
star of Bethlehem 126
Stellaria media 41
stephanotis 292
Stewartia ovata 219
stinking hellebore 37
stock, common 202
stork's bill, common 235
strawberry tree 311
strawberry, wild 157
Strelitzia reginae 316
Streptocarpus saxorum 294
striped squill 81
Strongylodon macrobotrys
    129
Styrax japonicus 137
summer hyacinth 358
summer snoflake 110
sun spurge 297
sunflower 220
superstitions *see also*
    fairies
    African lily 25
    angelica 174
    heather 320
    holly 267
    Lady's smock 153
    phlox 270
    rosemary 163
    snapdragon 254
    sweet pea 194
    winter-flowering
        honeysuckle 16
sweet alyssum 111
sweet box 353
sweet pea, wild 194
sweet violet 83
sweet William 207
sycamore 117
Sylvaticus, Mattheus 174
Syringa vulgaris 148
Syzygium aromaticum 368

Tagetes erecta 189
Tanacetum parthenium 233
Taraxacum officinale 322
Tatarian aster 332
tea plant 15
teasel, common 265
Tellima grandiflora 152

Texas bluebell 273
Thalictrum delavayi 251
Thiéry de Menonville,
    Nicolas-Joseph 277
thistle, spear 221
three-leaved cuckoo
    flower 73
Thunbergia alata 325
Thutmose III 80
Tillandsia ionantha 84
Titan arum 225
Tithonia diversifolia 338
toadflax, common 317
tobacco, wild 284
Trachelium caeruleum 141
Trachelospermum
    jasminoides 91
Trachystemon orientalis 46
trader's compass 91
Trifolium repens 86
Tropaeolum majus 47
tulip tree 161
Tulipa 'Semper Augustus'
    71
tulips, 'Semper Augustus'
    71
Turpin, Pierre Jean-
    François 366
Tussilago farfara 92
Tweedia coerulea 336
Tweedie, James 336
twinspur 48

Ulex europaeus 138
Urtica dioica 287

Vaccinium myrtillus 94
valerian 175
Valeriana officinalis 175
van Dyck, Anthony 220
van Gogh, Vincent 7,
    54–5
van Thielen, Jan Philip
    70
Vanda sanderiana 271
vanilla 282
Vanilla planifolia 282
Veitch & Sons 359
verbena 313
Verbena bonariensis 313
vervain 311
vetch, common 340
viburnum 60
Viburnum
    V. × *bodnantense*
        'Dawn' 60
    V. *opulus* 145
    V. *tinus* 348
Vicia sativa 340
Victoria amazonica 199
Victoria, Queen 199, 320
Viola odorata 83

violet
    dog's tooth 82
    sweet 83
Vitex agnus-castus 216

waling-waling 271
wallflower 79
    'Bowles Mauve' 61
Wang Gai 29–30
water avens 259
water lily
    banana 245
    giant 199
Waterhouse, John
    William 118
weeping forsythia 76
Welsh poppy 227
white clover 86
white dead-nettle 329
white forsythia 56
white fringetree 139
white gaura 308
white ginger lily 355
Wilde, Oscar 289
William of York 207
William the Conqueror
    207
Wills, William Gorman
    163
Wilson, Ernest 60
winter aconite 50
winter daphne 376
winter-flowering cherry
    346
winter-flowering
    honeysuckle 16
winter heliotrope 375
winter jasmine 24
winter winderflower 68
wintersweet 29
wisteria 142
wisteria sinensis 142
witch hazel 349
    orange 26
witches 16, 174
Withania somnifera 334
Wolffia globosa 274
wolfsbane 210
Wordsworth, William 72

yarrow 168
yellow corydalis 150
yellow rattle 266–7
yellow sage 18
Yonge, Charlotte M. 256

Zaluzianskya ovata 20
Zantedeschia aethiopica
    377
Zinnia elegans 291

PICTURE CREDITS

Acknowledgements

Firstly, I would like to thank the huge number of horticulturists, botanists, researchers and historians whose gathered information on plants over the past thousands of years not only made this book possible, but the world, for the most part, a hugely better place. Secondly, I would like to thank the Royal Botanic Gardens, Kew, where my love and understanding of plants was cultivated. Between the gardens, glasshouses and herbarium library, I spent some of my happiest years, growing plants, studying and researching. Thanks is also due to the Courtauld Institute of Art where I started my journey exploring art, beauty and the power of symbols.

The patience, love and support of my partner, the plantsman Andrew Luke must be acknowledged, for whom I am grateful to share our passion for the subject and who continues to support my learning.

Thanks is due to Tina Persaud for commissioning this book, as well as Kristy Richardson for her picture research. Encouragement in the pursuit of learning is down to my mother, father and brother, and thanks is due to Lucy Hall for her support and delight for me to write this book, as well as Jennifer Brown who kindly helped proofread. I am also indebted to the many friends including Aaron Bertelsen who continually inspire and keep the delight and enjoyment of plants wonderfully alive for me and many others.

And finally, I am thankful to my wonderful daughter, Lily Elizabeth Luke, who spent every moment in my company as I wrote this book. While this was down to biology rather than choice, I am grateful that she patiently waited for the book to be completed before being born.